ERATOSTHENES'

Geography

ERATOSTHENES'

Geography

FRAGMENTS COLLECTED AND
TRANSLATED, WITH COMMENTARY
AND ADDITIONAL MATERIAL, BY

DUANE W. ROLLER

PRINCETON UNIVERSITY PRESS

PRINCETON AND OXFORD

Copyright © 2010 by Princeton University Press
Published by Princeton University Press, 41 William Street, Princeton, New Jersey 08540
In the United Kingdom: Princeton University Press, 6 Oxford Street, Woodstock, Oxfordshire OX20 1TW

Library of Congress Cataloging-in-Publication Data
Eratosthenes.
    [Geography. English. Selections]
    Eratosthenes' Geography / fragments collected and translated, with commentary and additional material, by Duane W. Roller.
        p.   cm.
    Includes bibliographical references and index.
    ISBN 978-0-691-14267-8 (hardcover : alk. paper)   1. Geography, Ancient—Early works to 1800.   2. Physical geography—Early works to 1800.   I. Roller, Duane W.   II. Strabo. Geography.   III. Title.   IV. Title: Geography.
    G87.E6 2010
    913—dc22   2009022155

British Library Cataloging-in-Publication Data is available
This book has been composed in New Century School Book
Printed on acid-free paper. ∞
press.princeton.edu
Printed in the United States of America

10 9 8 7 6 5 4 3 2 1

Contents

Illustrations

Introduction

Interest in the size, shape, and inhabitants of the surface of the earth goes back to prehistoric times, as early humans moved beyond the limits of their own environment and encountered a world that was different from their own. The earliest literature is replete with travelers. Enkidu traveled to Uruk to meet Gilgamesh, Cain went from Eden to Nod, and Odysseus came to the Land of the Lotos Eaters. A primitive sense of geographical curiosity was an inevitable by-product of these wanderings. Enkidu met peoples whose lifestyles were different from his own, and Odysseus unfortunately learned both about the perils of sea travel and the dietary habits of the Lotos Eaters. The world was a complex place, and one's own clan was an insignificant part of its diversity.

Yet simple knowledge or even deep interest in the surface of the earth—whether its physical or anthropological qualities—did not automatically mean the development of geography. Scientific explanations for the character of the earth did not occur until the beginnings of Greek intellectualism in the sixth century BC; the Ionian monists Thales and Anaximandros were the first to theorize, however rudimentarily, about why the earth was the way it was. Yet only with Plato and Aristotle was there significant movement toward a discipline of geography, to be further stimulated by the extensive travels of Alexander the Great. But it was not until the efforts of the polymath Eratosthenes of Kyrene (ca. 285–205 BC), Librarian at Alexandria and tutor to the future King Ptolemaios IV, that geography took its place among the legitimate scholarly endeavors: indeed, it was Eratosthenes who created its terminology, including the very word *geographia* itself.

At some time during the 40 years after 245 BC, Eratosthenes wrote his three-book *Geographika*, the first scholarly treatise on the topic. Building on the thoughts of the previous three centuries, as well as the vast amount of data about places and peoples that had accumulated over the years, he laid out his conception of the nature of the surface of the earth (he had already determined its size in a previous treatise), with special attention to the *oikoumene*, or inhabited portions, and the peoples living therein. Topics as diverse as the depths of the seas, the geological history of the earth, its climate, and the customs of its population were included. This seminal treatise established geography as an academic discipline, and spawned numerous followers who themselves refined Eratosthenes' thoughts, among them Hipparchos, Polybios, Poseidonios, and Strabo.

Yet, as is so often the case with Hellenistic academic works, Eratosthenes' *Geographika* did not last long. Strabo's detailed summation from the Augustan period hastened its disappearance, and it is probable that the work was lost by the second century AC. This has placed the modern scholar in a difficult position: access to the thoughts of Eratosthenes means starting with Strabo (and the handful of other authors who quoted the work, none nearly as extensively) and working through his Roman perspective. Strabo's synthesis also carries the weight of all those who wrote on geography between the time of Eratosthenes and his own era. As is the case with all ancient texts that survive only through quotation by later authors, recovery of the original is a difficult process, for the quotations may have been chosen less to preserve the original author's thoughts than to fit into the agenda of the later source. A complex process of reverse recension is necessary to unravel Eratosthenes' ideas from those of Strabo and the other authors quoted by Strabo.

Since there was no existing text of the *Geographika* of Eratosthenes, there were no early modern attempts to reconstruct it, and it was not until 1789 that Günther Carl Fridrich Seidel attempted to pull the extant fragments from those who preserved them. The only other edition was that by Ernst Hugo Berger in 1880. Although both of these editions are significant, much has been learned since the time of Berger about reconstruction of texts from fragments, ancient geography, and the Hellenistic world. Moreover, the text of Strabo is in much better shape than it was in 1880, due primarily to the efforts of Wolfgang Aly and the editors of the ongoing Budé edition.

The present edition is not only the first in over a century, but the first with an English translation of the fragments. It builds on the present author's previous work on ancient geography, and one dares to hope that it will assist in modern comprehension of not only the origins of ancient geographical scholarship but the importance of Eratosthenes in ancient intellectual creativity. Berger's 1880 edition was the model, but a new selection of the fragments has been made, including some material not used by him, and ordering and book attribution has been refined. His original fragment numbers appear in parentheses after the new fragment number, yet the exact extents of Berger's fragments and of the present ones do not always correspond. In the analytical sections of this work, any fragment number without attribution (e.g. F1, F15) is from Eratosthenes; the authorship of other fragments is always stated (e.g. Dikaiarchos F121).

This edition of the *Geographika* also includes a set of maps, expertly drawn by the Ancient World Mapping Center at the University of North Carolina, that show virtually all of the over 400 toponyms cited by Eratosthenes. It is believed that this is the first time such an effort has been attempted. Although it is still in dispute whether Eratosthenes himself included a map with his text, these maps allow the modern reader to see the full spread of toponyms, from the northern Atlantic Ocean to beyond India, that were used to create the *Geographika*.

The author would like to thank the Loeb Classical Library Foundation of Harvard University, whose generous grant support allowed the completion of this project, as well as the Harvard College Library and the Library of the Ohio State University (especially its interlibrary loan services), James S. Romm and Grant Parker, both of whom read the manuscript and offered many valuable suggestions, Richard Talbert and the expert cartographers at the Ancient World Mapping Center, who produced the maps, and editor Rob Tempio, Natalie Baan, and many others from Princeton University Press. Further thanks go to A. B. Bosworth, David Braund, Stanley Burstein, John M. Fossey, Georgia L. Irby-Massie, and Letitia K. Roller.

Abbreviations

AClass: *Acta Classica.*

AFM: *Annali della falcoltà di lettere e filosofia, Università di Macerata.*

AHES: *Archive for the History of Exact Sciences.*

AJA: *American Journal of Archaeology.*

AJP: *American Journal of Philology.*

AntCl: *L'antiquité classique.*

BA: *Barrington Atlas of the Greek and Roman World* (ed. Richard J. A. Talbert, Princeton 2000).

Berger: Ernst Hugo Berger, *Die Geographischen Fragmente des Eratosthenes* (Leipzig 1880).

BiblArch: *The Biblical Archaeologist.*

BNP: *Brill's New Pauly* (Leiden 2002–).

Bosworth, *Commentary*: A. B. Bosworth, *A Historical Commentary on Arrian's History of Alexander* (Oxford 1980–).

BSOAS: *Bulletin of the School of Oriental and African Studies.*

CAH: The Cambridge Ancient History.

ClMed: *Classica et Mediaevalia.*

CQ: *Classical Quarterly.*

DGT: *Drevnojsije Gosudarstva na territori SSSR.*

Dicks, *Hipparchus*: D. R. Dicks, *The Geographical Fragments of Hipparchus* (London 1960).

DSB: *Dictionary of Scientific Biography.*

FGrHist: Felix Jacoby, *Die Fragmente der griechischen Historiker* (Leiden 1923–).

Fraser, *PA*: P. M. Fraser, *Ptolemaic Alexandria* (Oxford 1972).

Fraser, *PBA*: P. M. Fraser, "Eratosthenes of Cyrene," *PBA* 56 (1970) 175–207.

G&R: *Greece and Rome.*

Geus, *Eratosthenes*: Klaus Geus, *Eratosthenes von Kyrene* (Munich 2002).

GGM: *Geographi graeci minores* (ed. C. Müller, Paris 1855–82).

GRBS: *Greek, Roman and Byzantine Studies.*

HM: *Historia Mathematica.*

HSCP: *Harvard Studies in Classical Philology.*

IrAnt: *Iranica antiqua.*

JEA: *Journal of Egyptian Archaeology.*

JHS: *Journal of Hellenic Studies.*

Kidd, *Commentary*: I. G. Kidd, *Posidonius* 2: *The Commentary* (Cambridge 1988).

LSJ: *A Greek-English Lexicon* (ninth edition reprinted, Oxford 1977, ed. Henry George Liddell, Robert Scott, and Henry Stuart Jones).

NP: *Der Neue Pauly.*

OT: *Orbis Terrarum.*

OxyPap: Bernard P. Grenfell and Arthur S. Hunt, *The Oxyrhynchus Papyri* (London 1898–).

PBA: *Proceedings of the British Academy.*

PP: *La parola del passato.*

PSAS: *Proceedings of the Seminar for Arabian Studies.*

QJRAS: *Quarterly Journal of the Royal Astronomical Society.*

RE: Pauly-Wissowa, *Real-Encyclopädie der classischen Altertumswissenschaft.*

RÉG: *Revue des études greques.*

Roller, *Pillars*: Duane W. Roller, *Through the Pillars of Herakles* (London 2006).

RSO: *Rivista degli studi orientali.*

TAPA: *Transactions of the American Philological Association.*

VDI: *Vestiik Drevnii Istorii.*

Walbank, *Commentary*: F. W. Walbank, *A Historical Commentary on Polybius* (Oxford 1970–79).

ERATOSTHENES'

Geography

Eratosthenes and the History of Geography

The Background

The discipline of geography began with Eratosthenes of Kyrene and the publication of his *Geographika* in the last third of the third century BC. Before that time there had been interest in the surface of the earth, its formative processes, and its shape and structure, but it was Eratosthenes who brought these divergent streams of thought and experience together to create a new scholarly discipline. He also devised the terminology to accompany his ideas, with the new words "geography" (γεωγραφία) and "geographer" (γεωγράφος), based on the verb γεωγραφέω, "to write [about] the earth."[1] Eratosthenes' treatise was titled Γεωγραφικά (*Geographika*),[2] and the word "geography" was probably created by analogy with terms such as γεωμετρέω, "to measure [or survey] land," which itself had evolved from a technique, as Herodotos saw it, to a scholarly discipline.[3] In writing his treatise Eratosthenes built on a

---

[1] These terms appear in extant Greek literature in the *Geography* of Strabo (1.1.1, 16), from the Augustan period, but it is clear that they originated two centuries earlier with Eratosthenes, whom Strabo was quoting. The earliest extant citation of the word γεωγραφία is by Philodemos of Gadara (*On Poetry* 5.5), who lived from about 110 to 40 BC and was in Italy after the 70s (Tiziano Dorandi, "Philodemus," *BNP* 11 [2007] 68–73). The text is fragmentary, although it is clear that Philodemos was arguing against the belief that poets needed scholarly knowledge, listing as unnecessary disciplines music, seamanship, and geometry, as well as geography. It is not clear whom Philodemos was arguing against: the previous citation preserved is Herakleides of Pontos, the scholar of the fourth century BC, who seems too early to have used the word "geography," and it is probable that a more recent source is missing.

[2] This seems the most probable title (see Strabo 2.1.1). Also used are Γεωγραφία (*Geography*), at Strabo 2.1.41, which seems less likely, and Γεωγραφούμενα, a variant of Γεωγραφίκα that appears only in scholia (F117, 145). See Berger 17; Fraser, *PA* vol. 2, p. 756.

[3] Herodotos 2.109; Plato, *Theaitetos* 17 (162e). An older word, κοσμογραφίη, "describing the world," was apparently the title of a work by Demokritos of Abdera (Diogenes Laertios 9.46), who wrote a geographical treatise about which nothing is known (Strabo 1.1.1). Despite its origins in the fifth century BC, this word did not remain in favor,

tradition of interest in the surface of the earth and its landforms and inhabitants that went back to the beginning of Greek intellectualism. He
called Homer the first geographer (F1), a casual application of the technical vocabulary of the new discipline, but also recognition of Homer's
role in early understanding of the inhabited earth. Homer's world was
astonishingly broad, with vague knowledge of the mountains of central
Europe, the peoples north of the Black Sea, and the upper Nile and
pygmy tribes. There are hints of the climatic realities of the far north,
where the Kimmerians never see the sun, but much less knowledge of
the west, with nothing beyond Sicily, and no sense of any overall concept
of the earth or its surface.[4]

When Greeks began to spread into the western Mediterranean in
the latter eighth century BC, they learned about these regions as well
as the overall shape of the sea and its single, western outlet. About
630 BC one Kolaios of Samos became the first Greek to go out into the
Atlantic and to gain access to the wealthy mineral resources of the
southwestern Iberian Peninsula.[5] Although this was encroaching on
Carthaginian territory, both wealth and topographical data flowed back
into the eastern Mediterranean. To the south, the establishment of
Kyrene at about the same time provided some knowledge about interior
Africa, as the city soon became an outlet for far-ranging trade, the north
end of routes that originated south of the Sahara.[6] Interest in circum-
navigating Africa and connecting the lands south of the Red Sea to those
beyond the Pillars of Herakles resulted in a number of expeditions, the
first at the time of the pharaoh Necho II (610–595 BC). Phoenicians,
Carthaginians, and even Persians made the attempt.[7] Also in the seventh century BC Greeks settled the Black Sea and became more aware
of the rivers and peoples to its north. To the east the various great em-
pires could provide information, especially after the rise of Persia in the

---

to be replaced by *geographia*, which has essentially the same meaning: Eratosthenes
probably wanted new terminology for his new ways of thinking. Diogenes Laertios' use
(9.48) of the title *Geographia* for a treatise by Demokritos is anachronistic, perhaps an
adjustment of his title *Kosmographia*.

[4] For a discussion of Homeric geography and its numerous toponyms, see J. Oliver
Thomson, *History of Ancient Geography* (Cambridge 1948) 19–27; Germaine Aujac, *Ératosthène de Cyrène: Le pionnier de géographie* (Paris 2001) 19–22. A collection of translated texts of early geographical writers appears in George Kish, *A Source Book in Geography* (Cambridge, Mass. 1978) 9–72.

[5] Herodotos 4.152.

[6] Herodotos 2.32–3.

[7] Herodotos 4.42–3; Roller, *Pillars* 23–6.

sixth century BC. Cyrus the Great had gone to his death among the Massagetai, east of the Caspian Sea, in 529 BC, and about 15 years later Dareios I commissioned a certain Skylax of Karyanda to sail down the Indos and to return to the Red Sea. His published report was probably the earliest Greek travel account: it survived to be used by Herodotos.[8]

By 500 BC all the topographical and ethnographic information resulting from the expansion of the Greek horizon was beginning to find literary expression. Soon after Skylax, perhaps by 500 BC, the term λογογράφος ("writer of stories") was applied to those who used the new medium of prose to record city histories, ethnographies, and topographical data.[9] Among the several who are known, the most significant is Hekataios of Miletos, a prominent political leader in his own city around 500 BC, who wrote a *Circuit of the Earth* in two books, which included a discussion of Europe and Asia: he was probably the first to see the world in terms of continents.[10] His toponymic range was amazingly widespread: well into the Iberian Peninsula, Keltic territory, Skythia, the Kaukasos and beyond, India, and Ethiopia.[11] The extant fragments are mostly toponyms, so it cannot easily be determined how much ethnography and geography were included, yet many of his ideas matured and became more accessible in the *Histories* of Herodotos, written half a century later.

Hekataios also made use of a recent technique that would become inseparably connected with the discipline of geography. This was mapmaking, said to have been an invention of Anaximandros of Miletos, an important figure in both the theoretical and practical origins of geography. He was an associate or disciple of his famous compatriot Thales, the originator of Greek intellectual thought, which places Anaximandros in the first half of the sixth century BC. Significant in the development of Greek natural science and cosmology, he is of interest in the history of geography not only as the first mapmaker but also as the first

---

[8] Herodotos 4.44. The surviving text under the name of Skylax is actually an account from the fourth century BC, and not the work of the sixth-century explorer.

[9] Thoukydides 1.21; see also Aristotle, *Rhetoric* 2.11.7.

[10] The large number of topographical entries from Hekataios (over 300) in the *Ethnika* of Stephanos of Byzantion are defined as "Hekataios in his *Europe*" or "*Asia*" (*FGrHist* #1), although one must be cautious about the late derivative source, since Stephanos probably did not realize that Hekataios believed in only two continents (see Lionel Pearson, *Early Ionian Historians* [Oxford 1939] 31, 62–6). By early in the fifth century BC, the theory of continents was well established (Pindar, *Pythian* 9.8; Aischylos, *Persians* 718).

[11] Hekataios (*FGrHist* #1) F45–58, 184–93, 296–9, 325–8.

to conceive of the shape of the earth, although, as is generally the case with early Ionian Greek thought, the sources are elusive. Herodotos did not mention Anaximandros by name but referred several times to maps, derisively complaining about the inaccuracies of those of his own day and recounting the tale of how the Spartans were unexpectedly convinced not to give aid to the Ionian Revolt, because a map shown to them by Aristagoras of Miletos demonstrated how far away Persia was. Herodotos also provided a source for creating such a map, since in his *Histories* this incident is followed immediately by a precise itinerary of the Royal Road from Sardis to Sousa, complete with distances, stopping points, and a few topographical features, a digression that intrudes into the account of the career of Aristagoras.[12] These comments by Herodotos are the earliest extant discussions of maps. Later authors, including Eratosthenes himself (F12), specifically associated Anaximandros with the origins of mapmaking.[13] Thus the technique originated in Miletos during the sixth century BC: by the following century Herodotos, the first to realize that maps could mislead as well as inform, offered his critique and complained that they tended to over-regularize the surface of the earth. Yet mapmaking was firmly established as one of the essential tools for the emergent discipline of geography.[14] By providing a visual overview, maps also made it possible to relate distant portions of the world to one another, a strikingly new way of looking at the earth. Herodotos, probably using data from Hekataios, was able to suggest that Egypt, Kilikia, Sinope, and the mouth of the Istros lay on the same line,[15] which, although a crude calculation (the mouth of the Istros is over 500 km. west of the longitude of Sinope), is the first attempt to create a meridian. He did not have a specific term for this concept but merely stated that places "lie opposite" (ἀντίη κέεται) to one another, a term probably long understood by seamen to connect points on opposite sides of the Mediterranean. Yet this sense of "lying opposite" survived, and Eratosthenes was to use the concept for places that were far apart but on the same latitude or longitude.[16] The Greek word for midday,

---

[12] Herodotos 4.42, 5.49–54.

[13] The word for map seems originally to have been πίναξ (Strabo 1.1.11, quoting Eratosthenes, F12), a "plank" or "drawing board" (Homer, *Odyssey* 12.67; *Iliad* 6.169). Diogenes Laertios (2.2) used περίμετρον (originally "circumference" [Herodotos 1.185] but having the sense of "outline") for Anaximandros' creation, specifically noting that it was "of the land and sea."

[14] O.A.W. Dilke, *Greek and Roman Maps* (Ithaca 1985) 22–31.

[15] Herodotos 2.34.

[16] For example, southern India and Meroë (F50).

μεσημβρία, became the term for a geographical meridian, since when it was noon at a given place, every other point where it was also noon was on the same meridian.[17]

Anaximandros' other great contribution to geography was his theorization about the very shape of the earth itself. Building on Thales' view of a world floating on the water that was the basic component of the cosmos,[18] Anaximandros believed that the earth was shaped like a column (presumably the surface of a drum), perhaps reflecting the monumental stone architecture that was beginning to spread through the Greek world in his day.[19] Although his concept was soon abandoned, it marks the first systematic attempt to explain the overall shape and form of the earth, which, like mapmaking, would become an essential part of the discipline of geography. Yet the concept of the earth as a disk or column drum had apparent flaws, especially as seamanship revealed both the curved surface of the earth and the visible changes in celestial phenomena as one went north or south. Although there were inventive attempts to explain these pieces of information within the concept of an earth-disk,[20] before long thoughts turned toward conceiving of the earth as a sphere. This seems a Pythagorean idea,[21] connected with the harmonic and mathematical perfection of the cosmos and the sphere.[22] Parmenides of Elea, active in the fifth century BC, is also said to have tinkered with the concept, and perhaps was the first to divide the earth into climate zones.[23] It is with Plato that there is the first extant and extensive description of this new perspective of the world, as well as another important idea, that its inhabited part was only a small portion of the entire earth.[24]

[17] Aristotle, *Meteorologika* 2.5 (with textual issues), 3.5.

[18] Aristotle, *On the Heavens* 2.13.

[19] Duane W. Roller, "Columns in Stone: Anaximandros' Conception of the World," *AntCl* 58 (1989) 185–9.

[20] Anaximenes of Miletos, the traditional successor to Anaximandros, suggested that the northern parts of the earth were higher and thus caused the sun to disappear (Aristotle, *Meteorologika* 2.1).

[21] This is not the place to discuss the complex distinctions between Pythagoras, the emigrant from Samos who established himself at Kroton around 530 BC, and Pythagoreanism, which as early as the time of Aristotle (*Metaphysics* 1.5.1; *On the Heavens* 2.13) was recognized as an intellectual movement separate from the historical personality.

[22] The sources are summarized in William Arthur Heidel, *The Frame of the Ancient Greek Maps* (New York 1937) 63–80.

[23] Diogenes Laertios 9.21–2; Strabo 2.2.2.

[24] Plato, *Phaidon* 109b–110a. The Pythagoreans may have been the first to suggest that there were other continents beyond those surrounding the Mediterranean (Diogenes Laertios 8.25–6).

Many of these ideas were formalized in the work of Plato's associate the mathematician Eudoxos of Knidos. In his *Circuit of the Earth*[25] he put forth the idea that the inhabited part of the earth was much longer east-west than north-south, which led to the natural conclusion that India could be reached by sailing west from the Pillars of Herakles.[26] He may also have been the first to estimate the circumference of the earth,[27] and divided its surface into latitudinal zones, probably using the word that Eratosthenes was later to adopt, κλίμα, originally meaning "slope."[28] Eudoxos does not seem to have carried the idea of *klimata* far, but it laid the groundwork for Eratosthenes' more precise conceptualization.

By the mid-fourth century BC, the accumulation of knowledge made it possible for Ephoros of Kyme to include world geography as a major section of his history, published shortly after 340 BC. In this treatise he divided the world into four ethnic sections and provided more information on the north than previously available, and also discussed scientific issues such as the tides.[29] Although the word "geography" was probably still not yet in use, Ephoros moved closer than anyone previously to creating an actual scholarly discipline.

The proliferation of topographical and ethnographic knowledge because of the travels of Alexander is well known. This expansion of horizons created a large amount of data, especially about the remote eastern parts of the world, which was made accessible by the published reports of those with him, many of which were used as primary sources by Eratosthenes. Although the focus of the era and successive generations was toward the east, the west was also more intensively explored: Pytheas of Massalia traveled to the British Isles and reached the Arctic and Baltic in the 320s BC,[30] and an unknown traveler went down the West African coast perhaps as far as the Senegal River sometime between 361 and 335 BC.[31]

Despite the sudden increase in topographical data, theoretical speculation was not neglected. Aristotle's thoughts on geography are little

---

[25] Agathemeros 1.2; Plutarch, *On Isis and Osiris* 6; Thomson (supra n. 4) 117–18.

[26] Aristotle, *Meteorologika* 2.5, *On the Heavens* 2.14.

[27] He may be the source for Aristotle's unattributed figure of 400,000 stadia (*On the Heavens* 2.14); Thomson (supra n. 4) 116.

[28] Strabo 9.1.1–2; Polybios 2.16.3, 7.6.1. Although the extant use of the word is no earlier than the second century BC, it is clear from Polybios and Strabo that it had an earlier ancestry. See Karlhaus Abel, "Zone," *RE Supp.* 14 (1974) 989–1052.

[29] Ephoros (*FGrHist* #70), F30a, 131–4.

[30] Roller, *Pillars* 57–91.

[31] Pseudo-Skylax 112.

known, but he provided the first extant figure for the circumference of the earth (400,000 stadia) and suggested that one could reach India by sailing west from the Pillars of Herakles.[32] He was also crucial in preserving and synthesizing the ideas of his predecessors, and his *Meteorologika* contains a certain amount of geographical material, but no title survives that seems to indicate a purely geographical treatise.[33] Yet his students and immediate successors, the last generations before Eratosthenes, were active in the discipline. Dikaiarchos of Messana wrote a geographical treatise, perhaps titled *Circuit of the Earth*, and created the main terrestrial parallel, making the east-west length of the inhabited world one and one-half times the north-south.[34] He also may have been responsible for calculating the circumference of the earth at 300,000 stadia and further refining the terrestrial zones.[35] He was the first to make use of the information from the Arctic supplied by Pytheas, which far expanded the geographical extent of observed data. Another member of the Aristotelian school, Straton of Lampsakos, examined questions about the formation of the seas.[36] With the theories of Dikaiarchos and Straton, the study of the earth had reached the point where Eratosthenes was able to pull all former thought together and use his own original mind to create the discipline of geography.

The Life of Eratosthenes

Although biographical data about Eratosthenes are limited, it is possible to reconstruct a broad outline of his career.[37] He was the son of Aglaos

[32] Aristotle, *On the Heavens* 2.14. On Aristotle's geography, see Germaine Aujac, "Les modes de representation du monde habité d'Aristote a Ptolémée," *AFM* 16 (1983) 14–19.

[33] Paul Moraux, *Les listes anciennes des ouvrages d'Aristote* (Louvain 1954). Geographical data may have appeared in other works, such as the *Governments of 158 Cities* and *On the Rising of the Nile* (perhaps actually a work of Theophrastos: see Moraux, pp. 253–4).

[34] Dikaiarchos F122; Paul T. Keyser, "The Geographical Work of Dikaiarchos," in *Dicaearchus of Messana* (ed. William W. Fortenbaugh and Eckhart Schütrumpf, Rutgers University Studies in Classical Humanities 10, New Brunswick 2001) 363–5.

[35] Kleomedes 1.5; Fraser, *PA* vol. 2, p. 598; Thomson (supra n. 4) 154.

[36] For Straton, infra, pp. 130–1.

[37] For the major ancient sources see infra, pp. 268–70. The best modern sources are Fraser, *PBA* 175–207; Fraser, *PA* vol. 1, pp. 525–34; G. Knaack, "Eratosthenes" (#4), RE 6 (1907) 358–88; D. R. Dicks, "Eratosthenes, *DSB* 4 (1971) 388–93; Jerker Blomqvist, "Alexandrian Science: The Case of Eratosthenes," in *Ethnicity in Hellenistic Egypt* (ed. Per Bilde, Aarhus 1992) 53–73; Geus, *Eratosthenes*; Alexander Jones, "Eratosthenes of Kurene," in *The Encyclopedia of Ancient Natural Scientists* (ed. Paul T. Keyser and

and was born in the mid-280s BC in Kyrene. Both his name and that of his father are rare, indicating humble origins and demonstrative of the upward mobility possible in the Hellenistic world.[38] Kyrene, founded by Greeks in the seventh century BC, had long existed as a prosperous and cosmopolitan outpost of Greek culture, lying between Egyptian and Carthaginian territory, and serving as the contact point between the Greek world and interior Africa.[39] The city controlled a vast territory, perhaps more than any Classical Greek state. It had a rich economy, based largely on the export of horses and silphium. Libyans had long been known to the Greeks for their excellent horsemanship,[40] and the exotic herb silphium (σίλφιον, *ferula tingitana*), almost mystical in its reputation, had been exported to mainland Greece as early as the time of Solon and had a wide variety of culinary and medicinal uses.[41] Kyrene was full of distinctive art and architecture and had a flourishing intellectual tradition: in the fifth century BC the engaging personality Aristippos had come to Athens to study with Sokrates, and, with his daughter Arete and her son Aristippos, developed the Kyrenaian school of thought.[42] Like most Greek states, the history of Kyrene is one of political instability, with the city eventually coming under sporadic Persian control and then that of Alexander the Great. Upon Alexander's death Ptolemaios I provided a new constitution, although there continued to be occasional revolts and independent periods. It was into this environment that Eratosthenes was born during the last years of Ptolemaios I.[43]

By the late 260s BC Eratosthenes had gone to Athens for study. He was impressed with the vigorous intellectual environment of the city, and mentioned among his teachers Ariston of Chios, Arkesilaos of Pitane, who had recently become director of the Academy, and Bion of

---

Georgia L. Irby-Massie, London 2008) 299–302. For sources that focus on Eratosthenes as a geographer, see infra, pp. 32–5.

[38] Blomqvist (supra n. 37) 58–60.

[39] A.H.M. Jones, *The Cities of the Eastern Roman Provinces* (Oxford 1937) 351–60.

[40] Pindar, *Pythian* 9.4; Sophokles, *Elektra* 727; Herodotos 4.170, 189.

[41] Solon F39; Theophrastos, *Research on Plants* 6.3; Andrew Dalby, *Food in the Ancient World from A to Z* (London 2003) 303–4.

[42] Strabo 17.3.22; Diogenes Laertios 2.65–86.

[43] The *Souda* date of Ol. 126 (276–3 BC) seems too late to assume any study with Zenon of Kition (see Strabo 1.2.2), who died around 260 BC: Ol. 124 or 125, each requiring the change of only a single letter, seems better (Fraser, *PBA* 175–6). Eratosthenes' lifetime falls within narrow limits: he must have been old enough to have studied with Zenon, and he must have survived the death of Arsinoë III in 204 BC, dying between the ages of 80 and 82 (*Souda*; Censorinus, *On the Birthday* 15.2; Loukianos, *Makrobioi* 27).

Borysthenes, whose views were eclectic. Other teachers, according to Strabo, included the little-known Apelles of Chios and the founder of Stoicism himself, Zenon of Kition, although any contact would have been brief since he died shortly after Eratosthenes' arrival.[44] Also part of his education would have been the mathematical training at the Academy implicit in his later work. The *Souda* adds two more teachers, Lysanias of Kyrene, a philologist and grammarian with a Homeric interest,[45] and Eratosthenes' compatriot the famous Kallimachos. Whether or not Kallimachos was one of his teachers, Eratosthenes would have had regular contact with the most prominent Kyrenaian intellectual of the previous generation. Their paths would have crossed numerous times, and they evidently had academic disputes.[46]

Eratosthenes' exposure to the varied philosophical schools of Athens led Strabo to condemn his lack of depth as a philosopher, someone who had learned only enough to see philosophy as an escape (παράβασις). The extensive nature of Eratosthenes' later scholarly endeavors partially supports this view: he was expert, but not *the* expert, in many things, and thus was called "Beta" ("Second") or "Pentathlos."[47] Like so many interdisciplinary scholars from ancient to modern times, he could be assailed from all sides.[48] Strabo's irritation that he did not pay due respect to Zenon further reveals the independence of Eratosthenes' outlook and endeavors.

Thus Eratosthenes' education emphasized philosophy and, to a lesser extent, mathematics, with perhaps some philological training. Lacking is any evidence for a geographical education, but this is not unexpected given that there was no such discipline. Yet the Greek literature that Eratosthenes studied was replete with references to far peoples and places. This was especially true of Homer, whom Eratosthenes would come to believe was the first geographer. Hekataios' *Circuit of the Earth*, with its list of distant places, may still have been available.[49] Aischylos, Herodotos, and others provided material that would coalesce into the data for geographical scholarship. Half a century before Eratosthenes' birth,

---

[44] Strabo 1.2.2; Diogenes Laertios 7.1–38, 160–4, 4.28–45, 4.46–57. For Eratosthenes' education see Giorgio Dragoni, "Introduzione allo studio della vita e delle opere di Eratosthene," *Physis* 17 (1975) 49–52.

[45] Athenaios 7.304b, 14.620c.

[46] F8-9; the dispute is also implied in F2 and 12.

[47] *Souda*, "Eratosthenes."

[48] Strabo 2.1.41.

[49] It was certainly extant as late as the mid-fourth century BC: see G. L. Barber, *The Historian Ephorus* (second edition, Chicago 1993) 118–19.

Ephoros had been the first to write on world geography.[50] Eratosthenes' upbringing in Kyrene exposed him to exotic contacts at one end of the Greek world, and it is especially interesting that one of his teachers, Bion, came from the other end, from far-off Borysthenes, the collective term for the cluster of Milesian settlements at the mouth of the river of the same name (the modern Dneiper) at the north end of the Black Sea, one of the most remote areas of Greek settlement.

Eratosthenes also was born into a world of expanding geographical knowledge. It had only been 40 years since the death of Alexander the Great. The environment of the Successors had recently produced works by Androsthenes, Nearchos, Onesikratos, Ptolemaios I, and others recounting their travels with Alexander. The early Seleukids commissioned their own investigations into the far eastern reaches of the known world. The reign of Ptolemaios II (285/3–246 BC) would be one of further exploration, especially up the Nile and into East Africa, creating a new series of published reports. Although formal training in geography was impossible, Eratosthenes' world overflowed with geographical data.

After perhaps 20 years studying and writing in Athens, a period for which there are no details, Eratosthenes' career was dramatically affected by changing events in Alexandria. In early 246 BC Ptolemaios II died. Before the end of January, his son Ptolemaios III was proclaimed king, a reign that would last until the winter of 222/1 BC. More distinguished than the king was his wife Berenike II, from Kyrene. She was the daughter of Magas, who was related to the Ptolemies by marriage and had proclaimed himself king of Kyrene when Ptolemaios II came to the throne, but who decided late in life that a reconciliation with the Ptolemies would be wise, and proposed that his daughter marry the heir-apparent Ptolemaios III. This eventually happened, but not before Berenike killed her mother's candidate for her husband, the Antigonid prince Demetrios, who had been caught in her mother's bed. Although legally the Kyrenaika reverted to Ptolemaic control, Berenike continued as virtual queen of her own territory.[51]

The quarter century of rule by Ptolemaios III and Berenike II saw both the maximum political extent of the Ptolemaic empire and the greatest flowering of Alexandria as an intellectual center. The establishment of the Library by Ptolemaios I and its stocking from the great collections of the day, including, allegedly, those of Euripides, Aristotle, and

---

[50] Supra, p. 6.

[51] Justin 26.3; Günther Hölbl, *A History of the Ptolemaic Empire* (tr. Tina Saavedra, London 2001) 45–6.

Theophrastos,[52] had created an outstanding intellectual environment that, in terms of formal structure, was unprecedented. Scholars such as Euclid and Aristarchos of Samos were among the first to flourish in Alexandria. The reign of Ptolemaios II saw the arrival of the poets Theokritos, Kallimachos, and Apollonios of Rhodes. The most distinguished academic post in Alexandria was that of Librarian, which often carried with it the position of royal tutor.[53] The organization of the Library was largely implemented by Theophrastos' student Demetrios of Phaleron, who brought both political and academic credentials and was available because of the collapse of his political career. Eventually he was caught up in the intrigues of the regime change at the death of Ptolemaios I and was expelled by his successor, dying soon thereafter from the bite of an asp.[54] Demetrios was succeeded by Zenodotos of Ephesos, probably the first to hold the actual title of Librarian. He was known for his recension of early Greek poetical texts, especially Homer.[55] He was followed about a decade later by Apollonios of Rhodes, who lasted into the rule of Ptolemaios III, but who was forced into retirement shortly after the accession of the new king, who summoned Eratosthenes from Athens to be his replacement.[56] The role of Kallimachos in this, who seems to have been ill-disposed toward Apollonios and never held the position of Librarian himself, is uncertain. Both Apollonios and Eratosthenes were his protégés, although it is unlikely that Eratosthenes was ever Kallimachos' formal student,[57] yet it appears that the one out of favor was replaced with the one more preferred, who was also a fellow-citizen and whose shorter, more occasional poetry was more appealing than the epic style of Apollonios.[58] But the accession of Ptolemaios III and, more importantly, the Kyrenaian queen Berenike, certainly tilted the court

[52] Athenaios 1.3.

[53] The title normally translated as Librarian is Προστάτης, a common Ptolemaic term for the head of an organization. Zenodotos was probably the first to hold the office: see Fraser, *PA* vol. 1, pp. 320–335, for the early history of the Library and its organization.

[54] Diogenes Laertios (5.78) implied that it was an accident, but Cicero in 54 BC (*Pro C. Rabirio Postumo Oratio* 23, an oration with strong Alexandrian connections) said it was suicide. It is intriguing that this should be yet another case of suicide-by-asp from Ptolemaic Egypt.

[55] Fraser, *PA* vol. 1, p. 450–1.

[56] The sequence of Librarians is preserved in *OxyPap* 1241.

[57] Fraser, *PA* vol. 2, p. 490.

[58] Fraser, *PA* vol. 1, pp. 331–2. Nevertheless Eratosthenes and Kallimachos eventually ended up at odds with one another, since Eratosthenes suggested several times in the *Geographika* (F2, 8, 9, 12) that Kallimachos' scholarship was deficient.

toward Kyrene, and the major Kyrenaian intellectual in Alexandria, Kallimachos, would have had a say in the filling of the position of Librarian. Eratosthenes, moving at about the age of 40 into this flourishing cultural environment with its Kyrenaian slant, had already gained a reputation as a broad scholar and creative personality. His earliest publications were perhaps on philosophy, such as the treatise *Platonikos*, probably a dialogue, but with certain mathematical ideas that reflected his own interests rather than any Platonic heritage.[59] He was gaining fame, however, not for his philosophical writings, which were somewhat derivative, or his mathematical speculations, but as a poet in the tradition of Kallimachos.[60] This certainly played a role in his appointment as Librarian, as did his reputation as a broadly learned scholar, a true *philologos*,[61] as well as his nationality. He commemorated his appointment as Librarian by writing a poem that not only adeptly offered a mathematical proof of how to double the cube but honored the regime, especially the heir-apparent, the future Ptolemaios IV, and that was set up on a votive column in Alexandria.[62]

The early years of Eratosthenes' tenure seem to have seen an emphasis on mathematics.[63] He became a close associate of Archimedes, who may have visited Alexandria at about this time. He sent unpublished material to Eratosthenes for comment, and acknowledged his help with effusive praise in his *Method of Mechanical Theorems*.[64] This would have given Eratosthenes great credibility as a mathematician and prepared him for the treatise that would lead him from mathematics to geography, his *On the Measurement of the Earth*,[65] in which he set forth his method of calculating the circumference of the earth, a feat so profound yet so simple that it remains today one of the most amazing pieces of ancient scholarship, treated as such since antiquity.[66] Although

[59] See the detailed examination of the treatise by Geus, *Eratosthenes* 141–94.

[60] Fraser, *PBA* 184.

[61] The word, in its meaning of "a learned person," probably comes from Athenian oratory (the earliest extant example is Aristotle, *Rhetoric* 2.23.11, although oddly applied negatively, about the Lakedaimonians). Eratosthenes was allegedly the first to apply it to himself (Suetonius, *Grammarians* 10).

[62] Fraser, *PBA* 185–6.

[63] In fact he may have been responsible for organizing the mathematical and scientific collections of the Library: see Germaine Aujac, "Ératosthène, premier éditeur de textes scientifiques?" *Pallas* 24 (1977) 3–24.

[64] Archimedes, *On the Method of Mechanical Theorems*, *Preface* (see infra, p. 270).

[65] For the fragments of the *Measurement*, see infra, pp. 263–7.

[66] See, for example, Pliny, *Natural History* 2.247. Egyptian intellectuals were also impressed, and Elephantine, the island in the Nile at Syene, was represented in hiero-

any great scientific accomplishment relies for the most part on the creative abilities of a great mind, Eratosthenes was assisted by the topographical realities of his physical environment. The Nile extends almost due south for 1,000 km. from Alexandria to Syene at the First Cataract: its east-west deviation from this line is no more than 200 km. The route had been carefully measured on the ground by surveyors, most recently in the time of Ptolemaios II,[67] and there were reports from farther south and along the Red Sea coast, all published by explorers of Ptolemaios II such as Philon, Pythagoras, Ariston, Bion, and Simias.[68] It also was known that at Syene the sun was directly overhead on the summer solstice, and a well in the city demonstrated that there was no shadow at this time of the year.[69] Others had attempted to calculate the circumference of the earth, including Eudoxos of Knidos, Aristotle, Dikaiarchos of Messana, and Archimedes:[70] Eratosthenes built on these previous attempts and used Euclidean geometry to determine a circumference of 252,000 stadia. His erroneous assumption of a purely spherical earth had little effect.[71]

After publication of the *Measurement*, Eratosthenes turned to his *Geographika*, a natural extension of the earlier work. Having calculated the size of the earth, he now described what was on it, and, to some extent, how it was formed. Date of publication would have been between the accession of Ptolemaios III in 246 BC and Eratosthenes' death approximately 40 years later. His seeming obliviousness to the Roman expansion onto the Greek mainland that began in 218 BC, despite his

---

glyphics from the time of Ptolemaios III by a symbol combining the protractor and plumb bob. See Erich Winter, "Weitere Beobachtungen zur 'Grammaire du Temple' in der Griechish-Römischen Zeit," in *Temple und Kult* (ed. Wolfgang Helck, Wiesbaden 1987) 72–5.

[67] A close reading of Martianus Capella's description of Eratosthenes' techniques (6.596–8; see infra, p. 266) may indicate that Ptolemaios III had the route resurveyed for Eratosthenes' benefit, the only suggestion that he received research support (Blomqvist [supra n. 37] 65).

[68] On the sources for these explorers, see Duane W. Roller, *The World of Juba II and Kleopatra Selene* (London 2003) 234–5.

[69] Strabo 17.1.48. Syene, modern Aswan, was essentially the border post at the southern edge of Ptolemaic territory. Its latitude is 24°1', in Hellenistic times about 25–35 km. north of the tropic.

[70] On these, supra, pp. 6–7, infra, pp. 141–2.

[71] Fraser, *PA* vol. 1, pp. 413–15. His original measurement may have been 250,000, adjusted to provide a number divisible by 60: see the edition of Geminos' *Phainomena* by James Evans and J. Lennart Berggren, pp. 211–12. Modern calculations make the earth 40,008 km. around the poles and 40,075 at the equator, a minute difference.

detailed discussion of the Illyrian region where it started (F143–6), may suggest that the work had been completed by that time. He also began developing a chronological theory, a concept that, in a way, was connected with his geographical concerns. His first work on this topic was the *Olympionikai*, on Olympic lists,[72] which led to a broader chronological treatise, the *Chronographiai*,[73] the first universal chronology of Greek history, from the Sack of Troy to the death of Alexander, calculated at 860 years.

One other work connected to Eratosthenes' geographical scholarship is his poem *Hermes*, a combination of mythological and didactic material that described the childhood of the god.[74] It contains a Platonic description of the universe, and, most notably, an account of the terrestrial zones, poetically rendering the material in F44–5 of the *Geographika*.[75] There is no specific date for the poem: thus it is unknown whether it was an anticipation of the *Geographika* or a poetic version of parts of it. It influenced both Cicero and Vergil,[76] a feature of the interest in Eratosthenes' geographical works during Late Republican Rome.

Although Eratosthenes wrote in many areas, he is most remembered today for his geography and his chronology, fields in which, ironically, his views were soon superseded by the evolution of knowledge. His geographical data were seen as obsolete by the time of Strabo, largely because of the rise of Rome and the opening up of the western Mediterranean to Greco-Roman culture, and his chronology, although the basis for all later dating of antiquity, soon became buried under the more extensive Roman and Christian information. As is the case with many scholars noted for their breadth, he had few students. The *Souda* named four, of which the most famous is Aristophanes of Byzantion, his successor as Librarian, famous for his studies of Greek poets. Mnaseus of Patara in Lykia was a geographer, Menander, probably of Ephesos, an historian, and there was an otherwise-unknown Aristis.[77]

---

[72] *FGrHist* #241, F4–8; Geus, *Eratosthenes* 323–32.

[73] *FGrHist* #241, F1–3; see also F9–15; Geus, *Eratosthenes* 313–22.

[74] Geus, *Eratosthenes* 110–28.

[75] *Hermes*, F16.

[76] Cicero, *de re publica* 6.21–2; Vergil, *Georgics* 1.231–58; see also James S. Romm, *The Edges of the Earth in Ancient Thought* (Princeton 1992) 128–9.

[77] Aristophanes of Byzantion was Librarian from the death of Eratosthenes to about 189/6 BC (Fraser, *PA* vol. 1, pp. 332–3, 459–61. He produced editions of Homer and Hesiod and was significant in canonizing the Greek poets. Mnaseus' geographical work seems to have been heavily mythological (Athenaios 8.346; Fraser, *PA* vol. 1, pp. 524–5), and a few fragments survive of the treatise of Menander of Ephesos on the kings of Tyre (*FGrHist* #783).

In all probability, the final work of Eratosthenes is his biography or eulogy of Arsinoë III, the wife of Ptolemaios IV. Ptolemaios III died around 221 BC, and the convulsions thereafter included the murder of his widow Berenike.[78] Ptolemaios IV married his sister Arsinoë III, and Eratosthenes became a respected advisor to the young queen, her companion at public events.[79] Neither of the new monarchs lasted long: Ptolemaios IV died in 204 BC, and the queen, who was to be regent for the six-year-old Ptolemaios V, was promptly murdered by those who wished to have more influence over the new king.[80] She was barely 30 years of age. The composition of the *Arsinoë* is the last datable event in Eratosthenes' life, presumably written immediately after her death in 204 BC, although details remain uncertain. He was around 80 and did not survive much longer. When he died he was buried in Alexandria rather than in his hometown, a fact lamented in an epitaph by Dionysios of Kyzikos.[81]

The Geographika *of Eratosthenes*

Eratosthenes' *Geographika* was a modest work, only three books in length.[82] It probably did not survive intact past the second century AC and exists today in 155 fragments, mostly from the geographical treatise of Strabo of Amaseia, written in the Augustan period. Strabo quoted Eratosthenes extensively in his own first and second books, and throughout his work, providing 105 of the extant fragments. The only other author to make frequent use of the *Geographika* was Pliny the Elder, with 16 fragments. Scattered sources into Byzantine times provide minor details: the latest is Tzetzes.

The earliest extant author to cite the treatise is Julius Caesar.[83] Yet it was used extensively in Hellenistic times, especially by Hipparchos and Polybios, and one of the difficulties in understanding Eratosthenes is that these writers, who were generally ill-disposed toward the *Geographika*, themselves survive today only as quotations in Strabo's *Geography*, which also offers its own criticisms of Eratosthenes' treatise.

---

[78] Hölbl (supra n. 51) 127–8; on the eulogy, see Geus, *Eratosthenes* 61–8.

[79] Athenaios 7.276 (the only extant citation of the *Arsinoë*).

[80] Polybios 15.25; Hölbl (supra n. 51) 134.

[81] *Greek Anthology* 7.78; the *Souda* biography records that he died at 80, refusing food because his eyes were weakening.

[82] Strabo specifically mentioned the first (1.3.23 = F19), second (1.4.1 = F25), and third (2.1.1 = F47) books, but no more.

[83] Caesar, *Gallic War* 6.24 (= F150).

Strabo was generally quite supportive of Eratosthenes' views (except on Homer) and vigorously defended him against Hipparchos, yet Strabo's vantage point of the Augustan period meant that he often did not comprehend the geographically more limited world of two centuries earlier.[84] Moreover, as is often the case with Strabo, it is difficult to untangle his own views, those of his primary source, and those of the intermediate critical writings that he also used. Strabo's tendency to use pronouns with no obvious antecedent adds to the problem. For example, after mentioning Eratosthenes at 1.3.22, he is not named again until 2.1.5 (nine pages later in the Aly edition), although most of the intervening material is actually from Eratosthenes, using verbs without specific subjects. Moreover, between the two occurrences of the name "Eratosthenes" seven other sources are cited. Specifically, one need look only at 1.4.5, where the subject of διαμαρτών is not the previously mentioned source, Pytheas (cited three times), or the one before that, Hipparchos (cited twice), or even the one before that, Herodotos, but actually Eratosthenes, all the way back at 1.3.22. Strabo also mingled material of varying date, both earlier and later than Eratosthenes, as at 2.1.4–5, where Hipparchos is quoted, from his *Against the Geography of Eratosthenes*, and then three earlier writers, Patrokles, Deimachos, and Megasthenes, all of the era of the Successors, who were used by both Hipparchos and Eratosthenes. Finally Eratosthenes himself is named. All this becomes difficult to separate out: Patrokles is a case in point. His report of exploration in the Caspian region, from the early third century BC, survives only in fragments preserved by Strabo. But it does not seem that Strabo consulted Patrokles directly, but through both Eratosthenes and Hipparchos, whose opinions of the explorer were exactly opposed. At 2.1.2 the source for Patrokles is Eratosthenes but at 2.1.4 it is Hipparchos, giving two distinct views: Patrokles is either "greatly to be trusted" or "unbelievable." Although Strabo generally sided with Eratosthenes, believing Patrokles reliable, this is not always obvious, and it demonstrates the care that must be taken in Strabo's handling of the material.

In the *Geographika* Eratosthenes mentioned by name over 20 previous and contemporary authors. Most of these are from the time of Alexander and later, in other words, less than a century before the treatise was written. In addition there are unspecified sources, such as the sailors who traveled the Alexandria-Rhodes route (F128), as well as some

---

[84] On Strabo's attitude toward Eratosthenes, see Johannes Engels, "Die strabonische Kulturgeographie," *OT* 4 (1998) 76–81, and E. Hönigmann, "Strabon von Amaseia" (#3), *RE* 2nd. ser. 4 (1931) 132–6.

slight evidence of autopsy. As always, Eratosthenes may not have named all his sources, and identification of others may have dropped out in the complex recension of the text. Information on the upper Nile was probably not limited to that supplied by the two sources named, Philon (F40) and Timosthenes (F134). The authors that Eratosthenes used for the west of Europe are generally not recorded, in part because so much of this section of the treatise had become superseded and was not preserved by Strabo. Ephoros of Kyme, whose history was published around or just after 340 BC and was the first to include a section on world geography, is not cited in the extant fragments, but it is difficult to imagine that this precedent-setting treatise was unknown to Eratosthenes. Indeed one can presume that he had available to him in Alexandria everything that had been written on geography. This is why autopsy plays a minor role: beyond his personal knowledge of Egypt (F15), which itself may have been scant, Eratosthenes recorded only two places that he had visited, Helike in Achaia (F139) and Rhodes (F128). His use of two Arkadian vernacular terms (F140–1) suggests personal knowledge of that region, perhaps obtained at the same time that he visited nearby Helike.[85] Unlike Herodotos, Eratosthenes, who worked in the world's finest library, was not interested in fieldwork.

The literary sources range from Homer to Archimedes. In the section of Book 1 on the history of geography, Eratosthenes cited Homer, Anaximandros, and Hekataios of Miletos (F1). Homer was critiqued at length. Anaximandros was of interest because he was the first to create a map (F12) and to conceive of the shape of the earth. Hekataios (F1, 12) wrote the earliest geographical work and was also involved in early mapmaking. The contribution of Herodotos to the development of geographical scholarship is obvious, although the one existing citation (F20) objects to his skepticism. Yet these early authors were probably of value not so much as sources but as an outline of the history of geography.

In addition to Herodotos, there are several other authors cited from the fifth century BC. Xanthos of Lydia wrote a *Lydiaka* that included some comments about the formative processes of the earth (F15). Damastes of Sigeion wrote a number of geographical and ethnographic works, which Eratosthenes used for the area of the Persian Gulf and Cyprus (F13, 130), although he did not find him very reliable. Demokritos of

[85] Although he recorded (F61) an epigram set up in the Temple of Asklepios at Pantikapaion (modern Kerch at the mouth of the Sea of Azov), it is better to assume that he came across it in a collection in Alexandria, unless one wants to assume a vast amount of undocumented travel to remote places.

Abdera is also quoted (F33), but for his method of argumentation rather than geography, although he wrote a *Kosmographia*. The geographical material in Aischylos' *Prometheus* may also have been of value (F8).

After these Classical authors, no source is specifically cited in the extant fragments until those from the time of Alexander.[86] Eratosthenes relied extensively on the large number of published reports of those traveling in his entourage. Aristoboulos of Kassandreia provided data about the Oxos River (F109). The naval commanders Nearchos and Onesikritos were informative about India and the Persian Gulf (F41, 68, 77, 94). Nearchos' lieutenant Androsthenes of Thasos, who also explored the Persian Gulf, was of particular interest (F94): in fact Eratosthenes (through Strabo) preserved the longest extant fragment of his treatise *Sailing along the Indian Coast*. A little-known Orthagoras, seemingly also under Nearchos' command, was also a source (F94), as well as the equally obscure Anaxikrates (F95).

Perhaps also a companion of Alexander was Megasthenes. He visited the court of Chandragupta at Pataliputra, probably between 320 and 305 BC, the earliest Greek to go that far east. The experience produced his *Indika*, the first detailed discussion of the region, which Eratosthenes used several times (F21, 67, 69, 73). Two anonymous itineraries were also of value: a *Record of Stopping Points* (F73), listing places across northern India, may have emanated from a source around Megasthenes, and an *Asiatic Stopping Points* (F78), from the Kaspian Gates to India, may have derived from the official survey of Alexander's expedition.[87] Other unspecified itineraries were also used.

Also from the latter fourth century BC, but not connected with Alexander, is Pytheas of Massalia, Eratosthenes' primary source for northwest Europe and beyond. Understanding his debt to Pytheas is difficult because Strabo (largely influenced by Polybios) was totally dismissive of the Massalian explorer and his epic journey to northwestern France, the British Isles, and beyond. Strabo probably discarded much of the information that Eratosthenes obtained from Pytheas, although occasional facts survived (F37, 132–3), often negatively presented by Strabo, or perhaps missed in his editing.

---

[86]There may be an allusion to Ephoros at F1 and to Theopompos at F8.

[87]The expedition was surveyed by Diognetos and Baiton: the latter wrote a *Stopping Points of Alexander's Expedition* (Pliny, *Natural History* 6.61; Athenaios 10.442b). See also the unspecified "published record of stopping points" in F73. There may have been a summary or synthesis of these itineraries in the Alexandria library (M. Cary and E. H. Warmington, *The Ancient Explorers* [revised edition, Baltimore 1963] 280; P. M. Fraser, *Cities of Alexander the Great* [Oxford 1996] 80–6).

Eratosthenes also used many sources from the generation after Alexander. Patrokles traveled into the Caspian Sea region and from there to India, probably in the 280s BC, although there are difficulties with his account because it presumes a Caspian Sea connected to the External Ocean and a sea route from it to India.[88] Nevertheless the report was quoted for the dimensions of India (F47, 50, 69). There was also Deimachos of Plataia, who was at the court at Pataliputra during the reign of King Bindusara (294–269 BC). Eratosthenes was dubious about the reliability of Deimachos' history of India (F22), not unexpectedly, since a surviving fragment describes anatomically unlikely people. Eratosthenes found him generally inferior to Megasthenes (F67).

In the early third century BC, Euhemeros of Messene, associated with the Makedonian king Kassandros (reigned 305–297 BC), wrote a fictionalized account of a voyage to unknown regions around India, titled *The Sacred Record*. Eratosthenes did not find him reliable (F8, 13), although Strabo implied that he was inconsistent in this assessment. Fantasy travel accounts, a genre that developed in Hellenistic times and continues into today, have much embedded in them that may be accurate, but it is difficult to analyze them properly, as they can easily be dismissed in their totality. There is no indication what Eratosthenes found useful in *The Sacred Record*.

Although Aristotle himself is not mentioned in the extant fragments of *Geographika*, some of his immediate successors were included, especially his student Dikaiarchos of Messana, the most recent writer on geography in Eratosthenes' day (F132). In addition, there was Straton of Lampsakos, who succeeded Theophrastos around 287 BC as head of the Academy and was also tutor to Ptolemaios II.[89] His research on the processes of nature was of interest to Eratosthenes, who had a high opinion of his scholarship (F15) and absorbed his theories about the level of the seas and siltation.

During the reign of Ptolemaios II (285/3–246 BC) a number of explorers were sent up the Nile and into the regions between the Nile and the Red Sea. Eratosthenes was probably familiar with all their reports, although he mentioned by name only Philon and Timosthenes (F40, 97, 134).[90] The former went as far south as Meroë and published an *Aithiopika*, and was interested in southern latitudes. Timosthenes explored beyond the mouth of the Red Sea and wrote a treatise titled *On Harbors*,

---

[88] Strabo 11.11.6; Pliny, *Natural History* 6.58.
[89] Diogenes Laertios 5.58–64.
[90] For Timosthenes, see *FGrHist* #670.

which seems to have included the entire circuit of the Mediterranean, but Eratosthenes was ambivalent about its value.

The latest source mentioned in the *Geographika* is Eratosthenes' colleague Archimedes. Almost exactly the same age, the two regularly exchanged ideas. Eratosthenes disagreed with Archimedes' view that any surface of water would be curved like the surface of the earth, noting that the level of the Mediterranean varied at different places (F16), but it is not certain whether his objections appeared in the *Geographika* or Strabo was merely citing divergent opinions of two academics.

There is no reason to believe that Eratosthenes named all his sources. Interest in the Italian connections of Odysseus may reflect the *Telegonia* of his compatriot Eugammon of Kyrene, probably of the sixth century BC. Eratosthenes' information on the West African coast (F107) is unattributed, although one presumes that it is based on one of the several existing sailing itineraries, such as that of Hanno (ca. 500 BC) or that known today as Pseudo-Skylax (mid-fourth century BC). What use was made of Ephoros' writings, essential to Eratosthenes' research, is uncertain (F1). Knowledge of Euclidean geometry pervades the treatise (e.g. F49, 51). Eratosthenes also used unspecified "ancient maps" (F51, 64) but rarely agreed with them. There were many itineraries from the era of Alexander and earlier that were of value. Moreover, despite his library facilities, published works were not his only source of information. Alexandria was a vibrant and cosmopolitan seaport city, and Eratosthenes would have had daily contact with people from all over the world.[91] He talked to seamen who went to Rhodes (F128) and doubtless others about their own sailing routes. The way overland to his hometown of Kyrene (F101), now a Ptolemaic possession, was well known. His comment that he used "distances that have been handed down" (F131), a methodology that earned him the criticism of Hipparchos but the sympathy of Strabo (F52), is further revealing about his technique. There is also a single hint that he made some of his own latitude calculations (F128). Some things not even the world's greatest library could provide.

From the beginning of his own *Geography* through much of the second book, Strabo provided a summary of Eratosthenes' work, mentioning him many times, and citing him frequently even when not specifically named. Despite the difficulties in deconstructing Strabo, the plan

---

[91] Although from a later period (early second century AC), Dio Chrysostom's *To the Alexandrians* (32), especially section 36, well sums up the cosmopolitan international nature of the city.

of Eratosthenes' treatise is clear. Yet because it exists today only in paraphrases and summaries, little can be said about his written style, although like many Hellenistic authors, he created new terminology to explain his new ideas.[92] The only lengthy direct quotation is F30 (which ironically does not mention Eratosthenes by name). The passage lays out the character of the surface of the earth in a straightforward fashion, presented in proper scientific style with propositions and proofs, and free of the ambiguities and layering of source material that characterize Strabo's text. It is quite extensive, and remains as the only certain lengthy passage of Eratosthenes' actual text, although there is also a direct quotation in F33, similar in style. In several other places Strabo may have been quoting directly, but Eratosthenes himself was following his own sources, and so the diction may not be that of Eratosthenes. At F74, on India, the wording is probably that of Megasthenes. At F94, on the Persian Gulf, Eratosthenes is introduced in indirect discourse but the account changes to direct, probably Eratosthenes' own quotation of Androsthenes and, later, Nearchos and Orthagoras. At F95, on Arabia, there is a similar change in discourse, but no source on Arabia proper is ever named, and this may in fact be Eratosthenes' own synthesis. All other fragments seem to be paraphrases.

It is also uncertain whether a map was included, although one can easily be created from the data.[93] The material contained in F25 and 30 is often assumed to be instructions for making a map, but there is no specific evidence here that Strabo was in fact looking at one (the key terms for map, such as πίναξ, are absent), or doing anything other than summarizing a literary text. A reference to a map of Eratosthenes in the scholia to the *Description of Greece* of Dionysios seems too late and isolated to be definitive.[94]

Book 1 of Eratosthenes' *Geographika* opens with the history of geography from the time of Homer. Eratosthenes was concerned with problems in Homeric geography, especially in the west. A disproportionate number of fragments (F2–11) examine the issue, probably because the topic was given extra emphasis by Strabo, who intensely disliked Eratosthenes' point of view, which Strabo saw as disrespectful to the originator of Greek literature and culture. There are hints of discussions of

[92] Infra, p. 23.

[93] See Thomson (supra n. 4) 135, 142, or the detailed plans in vol. 1 of the Aly edition of Strabo; see also Dilke (supra n. 14) 32–4; Heidel (supra n. 22) 122–8.

[94] Scholia to Dionysios 242 (*GGM* vol. 2, p. 441). Other fragments that have been suggested to refer to a map include F47, 56, and 72. See Bosworth, *Commentary* vol. 2, pp. 213–15.

other authors down to Herodotos. After this description of the earliest
sources for geographical knowledge, Eratosthenes examined the specific ones that he used for his treatise, from the sixth century BC to his
own era, stressing the great advances in geographical knowledge made
in the period of Alexander and the Successors (F15). His latest sources
are virtually of his own time, from the early Seleukid period (F50) and
that of Ptolemaios II (F40).

Eratosthenes then proceeded to examine the nature of the earth itself (the transition is apparent at F15), especially its shape and formative processes, noting the presence of marine phenomena far inland and
the effects of siltation on coastal regions and river estuaries. Much of
the data comes from Egypt and may be autopsic, although such thoughts
had long been current in Greek intellectualism, and Eratosthenes mentioned both Straton of Lampsakos and Xanthos of Lydia.

Book 1 closes with a section on fantasies and fabrications, although
the extant fragments are tangled with Strabo's own interpretations and
prejudices. Strabo placed within the realm of fabricated geography details that Eratosthenes found reliable, most notably the data from Pytheas of Massalia (F153). Eratosthenes' emphasis seems to have been
the stories about the early wanderings of Herakles and Dionysos,[95] and
perhaps other mythic material, as well as the fantastic tales that had
emanated from India since the time of Herodotos (F22). That this discussion closed Book 1 was made clear by Strabo (F19).

Book 2 is devoted to the shape of the earth. Analysis of this book is
not as straightforward as Books 1 and 3, because material from Eratosthenes' *On the Measurement of the Earth* is tangled into the extant
sources, and it is even possible that a summary of it was included in the
Geographika. The exact order of topics in Book 2 is not obvious. It may
have opened with comments about the actual shape of the earth (F25,
33) and some of the measuring techniques that were used in the calculations (F26, 27). Eratosthenes outlined his basic premises about the inhabited world (using the Aristotelian term οἰκουμένη),[96] believing that
it was longer east-west than north-south (F33), and emphasizing that
one could sail west from Spain to India.[97] As usual, Strabo, the extant

[95] Eratosthenes' rejection of the Indian mythology of Dionysos also appeared in his
Katasterismoi: see Jordi Pàmias, "Dionysus and Donkeys on the Streets of Alexandria:
Eratosthenes' Criticism of Ptolemaic Ideology," *HSCP* 102 (2004) 191–8.

[96] Aristotle, *Meteorologika* 2.5.

[97] Columbus was familiar with Eratosthenes, through a Latin translation of Strabo
(four editions appeared between 1469 and 1480) and Aeneas Sylvius Piccolomini's
Historia rerum ubique gestarum (1477), which summarized the data in Eratosthenes'

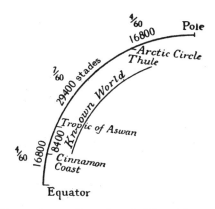

Figure 1. The Zones of Eratosthenes (based on J. Oliver Thomson, *History of Ancient Geography* [Cambridge 1948], fig. 19, and redrawn by Princeton University Press).

source, was following both Hipparchos' objections to these theories and their methodology as well as Eratosthenes himself. Discussion of the inhabited landmass led naturally to an examination of the External Ocean (F39), which may have included comments on the tides.

The bulk of Book 2 laid out Eratosthenes' conception of the earth, using the circumference of 252,000 stadia that he had determined in the *Measurement* (F28). Although specific terrestrial distances were largely reserved for Book 3, there were some calculations (F34). In addition, he divided the inhabited world latitudinally into five zones (F30, 45), devising inventive terminology to do so. The surface of the world around the *oikoumene* was a σπόνδυλος, a vertebra or spindle whorl (F30–1), in the middle of which lay the *oikoumene* itself, an island shaped like a chlamys (χλαμυδοειδής).[98] Thus Eratosthenes combined scientific and domestic vocabulary to create a striking description of the surface of the earth. All this is preserved in a lengthy passage by Strabo,[99] a rare case in which he quoted Eratosthenes directly without going through Hipparchos or others.

It is easy to shift much of the material that early modern editors included in Book 2 either to the *Measurement* or to Book 3, with mathematical calculations in the former and topographical details in the latter.

Measurement (see V. Frederick Rickey, "How Columbus Encountered America," *Mathematics Magazine* 65 [1992] 219–23).

[98]This was a word probably created by Eratosthenes that came to be applied to the city of Alexandria itself (Strabo 17.1.8); Klaus Zimmermann, "Eratosthenes *Chlamys*-Shaped World: A Misunderstood Metaphor," in *The Hellenistic World: New Perspectives* (ed. Daniel Ogden, London 2002) 23–40.

[99]Strabo 2.5.5–6 (= F30).

Yet to do this too vigorously strips Book 2 of a large amount of its potential detail. The complex mathematics of the *Measurement* was probably
not repeated, unless in brief summary form, but the results were essential in understanding the geographical conclusions. The matter of topographical detail and the terrestrial distances is somewhat more inscrutable. Toponyms were necessary as Eratosthenes set forth his view of
the extent of the world, and not every toponym belongs in Book 3. Yet
one would expect as much condensation as possible (despite a complaint
by Strabo about lack of brevity),[100] not to interfere with the clarity of
conceptual issues. It is possible, however, that there was a digression on
Syene (F40–3), the topographical nexus of Eratosthenes' calculations
and theoretical structure: its well and the lack of shadows in midsummer
became a mythic part of world topography and geography, and it retained
a unique importance as the place from which all measurements originated, where, as Pliny wrote, "the world was grasped."[101]

In Book 3 Eratosthenes described in topographical fashion the inhabited world. More fragments survive from this book because of the
vast amount of topographical detail that was useful to others. He broke
with the pattern accepted since Hekataios and did not go in clockwise
fashion from the Pillars of Herakles but from east to west (F78), perhaps reflecting the contemporary obsession with India. Rather than see
the world as populated by ethnic groups, as was the case as recently as
the time of Ephoros, Eratosthenes emphasized places, whether cities or
geographical features, reflecting the new attitudes of the early Hellenistic world.[102] The book opens with the establishment of two baselines, an
east-west one from the Pillars of Herakles to India and a north-south
one from Meroë to Thoule (F64–5), with the two crossing on Rhodes.
This provided a method by which Eratosthenes could divide the world
into sections. He also discussed some of his sources (F51), which were
written and oral reports of distances.

Eratosthenes then listed his parallels. The normal way of calculating
latitude (and hence, parallels) was by length of longest day (F59–60), a
methodology already used by Pytheas, if not earlier, but sailors' and travelers' reports also played a role, and Eratosthenes knew that his parallels and meridians were rarely straight lines (F131). Starting in the
south with the Kinnamomophoroi territory (the horn of Africa), the primary parallels were at Meroë, Syene, Alexandria, Rhodes, Lysimacheia

---

[100] Strabo 1.4.1 (= F25).

[101] Pliny, *Natural History* 6.171 (= F42): "mundo ibi deprehenso."

[102] Klaus Geus, "Space and Geography," in *A Companion to the Hellenistic World* (ed.
Andrew Erskine, paperback edition, Oxford 2005) 243–4.

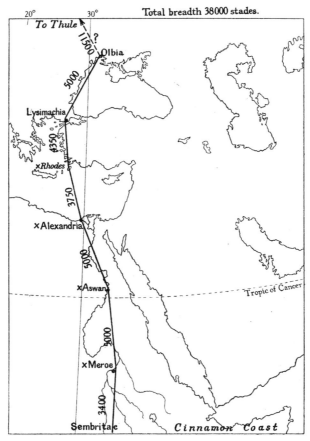

Figure 2. The Prime Longitude Line of Eratosthenes (based on J. Oliver Thomson, *History of Ancient Geography* [Cambridge 1948], fig. 20, and redrawn by Princeton University Press).

in Thrace, Borysthenes, and Thoule (F60). Extension of these parallels to the east and west was one of the more problematic issues: Eratosthenes was criticized by Hipparchos and others for using travelers' reports rather than astronomical data, since as far as possible his parallels and meridians connected known points, often at the expense of being straight lines. There is an obvious lack of data for some of the northern parallels (Europe in the west and central Asia in the east) and the western extensions of the southern ones (interior Africa).

Next were the meridians. To the east of the base meridian through Alexandria and Rhodes were those through Thapsakos on the Euphrates, the Kaspian Gates, and India, and in the west through Messana and the Pillars of Herakles (F62–5, 80, 92). The courses of the meridians, especially in the east, were not delineated with certainty. Continuing his use of inventive terminology, Eratosthenes conceived of dividing

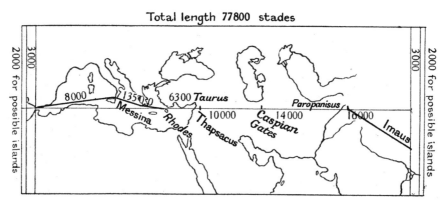

Figure 3. The Prime East-West Parallel of Eratosthenes (based on J. Oliver Thomson, *History of Ancient Geography* [Cambridge 1948], fig. 21, and redrawn by Princeton University Press).

the world into portions that he called σφραγῖδες (gem or seal stones). This was by no means a neologism, having existed since at least the fifth century BC, but was newly used in Hellenistic Egypt to describe a surveyed and numbered plot of land.[103] Eratosthenes, with his taste for innovative vocabulary, took the word in this meaning, and gave it a still newer sense as a broad geographical division (F66).[104] The precise shape that he had in mind is unclear, perhaps a rectangle with rounded corners. As a geometrical form, the concept reflects the influence of Euclid.[105] Yet the analogy proved untenable and Eratosthenes seems to have applied it rigorously only to the eastern regions (India, Ariana, and Mesopotamia), with serious problems apparent in Mesopotamia. There is a single reference to a fourth peculiarly shaped sealstone of Arabia, Egypt, and all Aithiopia (F92), but further discussions of these regions do not use the term and sealstones do not appear farther west or in the north. It is probable that he devised them to apply to India and Ariana but wisely dropped the term when discussing regions in and around the Mediterranean. No later geographer revived the term, but

[103]Herodotos 1.195; see the several papyrus citations in *LSJ*, although these are slightly later than Eratosthenes; E. H. Bunbury, *A History of Ancient Geography* (second edition, London 1883) vol. 1, pp. 654–5; A. Thalamas, *Le géographie d'Ératosthène* (Versailles 1921) 241–7.

[104]That the word was used in contemporary Ptolemaic land division has been the general opinion for over a century, but recently has been disputed by Klaus Geus, "Measuring the Earth and the *Oikoumene*," in *Space in the Roman World* (ed. Richard Talbert and Kai Brodersen, Münster 2004) 20–4.

[105]Katherine Clarke, *Between Geography and History: Hellenistic Constructions of the Roman World* (Oxford 1999) 207.

the idea represents a significant advance in geographical theory, moving from a world defined ethnically to a purer geographical conception based on landforms.

The rest of the treatise is an examination of the inhabited world, from east to west. In terms of the fragments, India received the most attention, perhaps expected in the century after Alexander, but survival of this material is also due to the interests of Strabo's own day, when Augustus attempted to strengthen trade between India and the Roman world.[106] Only in the Indian section is there an extant discussion of problems in analyzing the sources for a specific region (F50, 67–8). There is also an examination of the routes between the Mediterranean and India (F108–9), an interest reflecting the world of Alexander and the Successors, but not apparent elsewhere in the treatise. How much, if any, ethnographic material was included is uncertain, given a tendency (reinforced by Strabo) to dismiss it as fantasy, although Eratosthenes' use of Megasthenes and other ethnographic writers and their cultural data is apparent. These sources provided an itinerary across India all the way to the mouth of the Ganges (F69) and detail about the climate and economy of the Indos region (F74). Taprobane was the most remote place in this direction, still only vaguely known when Eratosthenes wrote.

From India the discussion moved to the second sealstone, Ariana. The extant comments are sparse, and the description merges into the third sealstone, Mesopotamia (F83). Information about Ariana is limited to basic geographical features such as rivers, mountains, and ethnic groups. As Strabo presented it, Eratosthenes' main emphasis was on the boundaries of the sealstone.

The third sealstone is Mesopotamia. Again the surviving fragments focus on the boundaries and topographical features, although there is ethnographic detail, such as the asphalt production of Babylonia (F90) or the peculiar characteristics of Lake Thopitis (F87). Eratosthenes also provided a summary of Androsthenes' account of the coasts of the Persian Gulf (F94). It is with Mesopotamia—and its physical relationship to Ariana—that problems begin to appear with the sealstones. Eratosthenes had difficulty in fitting known data about the boundaries, river courses, and main routes into his scheme, and there are some major inconsistencies, especially in the northern portions. The fourth sealstone follows almost seamlessly with a lengthy discussion of the Arabian peninsula (F95). No source for this is cited in Strabo's recension, yet the emphasis on trade items suggests that Eratosthenes probably relied on

---

[106] Strabo 17.1.13, 2.5.12; Augustus, *Res gestae* 31.

merchants in Alexandria, with additional data from sailing records. The account of the Persian Gulf and Arabia shows more than anywhere else that ethnography and natural history were a significant part of the work, but that much of this material has not survived, eventually seen as derivative, fantastic, or obsolete.

Theoretically Egypt was part of the fourth sealstone, although the concept was ignored in favor of a general view of Africa as a whole. Whether the detailed information on Syene (F40–3) belongs here or in Book 2 is uncertain. But there was an analysis of the course of the Nile (F98) as far south as Meroë as well as information on the territory between the Nile and the Red Sea. Like so many before him, Eratosthenes was intrigued by the flooding of the Nile (F99). West of Egypt, his interest spread across coastal North Africa to the Atlantic coast and Lixos (F107) and Kerne (F13), the latter at the mouth of the Senegal and the most remote toponym in this direction. The surviving information is limited. There are some hints of coastal details (F107) yet nothing about interior Africa west of the Nile, although the dimensions of the continent (hardly known in the south) were of interest.

Eratosthenes then described the northeastern parts of the inhabited world, from the Caspian Sea to the Pontos (Black Sea). The extant comments are all geographical with no ethnography surviving. He believed, in the fashion of his day, that the Caspian Sea was an inlet of the External Ocean (F110). His data on the Pontos reflect a sailing itinerary (F114). The following region, Anatolia and the Greek islands, has little detail, even of a geographical nature, and most of the fragments concern distances, although there is a discussion of mining on Cyprus (F130), an early report of the effects of industry on the landscape. There are also comments about the Mediterranean itself, concerned with distances, its extremities, and component parts (F127–9).

The remainder of the treatise examines the north coast of the Mediterranean, in other words, Europe. This is the region where geographical knowledge had advanced the most between the time of Eratosthenes and Strabo, and Strabo was in no way hesitant about pointing out his predecessor's ignorance, although he did realize that the region had been alien in Eratosthenes' day.[107] Eratosthenes described how the southern edge of Europe was divided into "promontories" (Greece, Italy, and Iberia, F135) with, expectedly, more detail about the Greek peninsula than the other two. There are comments about Illyria (F143–6) and the upper Adriatic into the Danube region (F13, 148–9). Julius Caesar

[107] Strabo 2.1.41, 2.4.1–2.

was aware that Eratosthenes knew about central Europe (F150), yet there is scant material on Italy proper (F151), where the obsolete account probably did not survive into the Roman period.

The final region is western Europe. Here, Strabo felt, Eratosthenes had ventured into the area of the fabulous (F13), because his primary source was Pytheas of Massalia (F153), whom Strabo refused to believe had any reliability whatsoever. Moreover, Strabo did not use Eratosthenes directly for his assessment of Pytheas, but Polybios, who had his own highly negative views. Eratosthenes, however, felt that Pytheas had much to say about northwest Europe, the British Isles, and beyond. Strabo's statement about Eratosthenes' ignorance of Germanika and Brittanika (F131) reveals that he probably did discuss those regions. Other comments about the outer coast of the Iberian Peninsula (F152–3) show inclusion of that area. Yet the vast amount of data from Pytheas' *On the Ocean*, including material on the islands around and beyond Britain, as far as Thoule, was generally mentioned by Strabo only to reject the veracity of Pytheas, not to critique Eratosthenes. Thus it is far from clear how Eratosthenes processed Pytheas' data. He was familiar with *On the Ocean*, since he could have learned about Thoule from no other source (F34, 37), and the appearance of a number of Keltic toponyms also suggests Pytheas, but all were rejected by Strabo as fantasies. There is no evidence as to what use, if any, Eratosthenes made of Pytheas' information on the Baltic and the rivers of northern Europe, although Strabo, in his typically engaging way of providing detail about a source that he vigorously rejected, knew that Pytheas had been in the Baltic.[108]

To bring the *Geographika* to a close Eratosthenes considered issues of ethnicity (F155), using the model of virtue of his own era, Alexander the Great, who allegedly had rejected the traditional division between Hellenes and barbarians and its implicit assumption that the former were good and the latter bad. Alexander had been advised to look at humanity in a more orthodox way but refused to do so, choosing rather to distinguish people in terms of individual virtue rather than ethnicity. Eratosthenes may have originally developed his thoughts in his own essay on Alexander, building on concepts learned from his teacher Zenon of Kition,[109] eventually to place them at the end of the *Geographika*.

---

[108] Strabo 1.4.3. One of Strabo's more valuable techniques is to describe material in great and loving detail and then reject it, often contemptuously: this is most apparent in the fascinating tale of Eudoxos of Kyzikos (Strabo 2.3.4–5).

[109] Plutarch, *On the Fortune and Virtue of Alexander* 6–8, is a fuller rendering of the thought of F155, attributed to the *Republic* of Zenon of Kition as well as Eratosthenes'

The comments about virtue and ethnicity are a suitable close to the treatise, not only reflecting the complex ethnic interaction of the Greek world that had begun in Alexander's day and was continuing in Eratosthenes' time, but a reinforcement of the ethnic diversity of the *Geographika* itself, extending from the Indians to the Anariakoi northeast of the Caspian Sea, the nomads of North Africa, and the Kelts and British. Such a work would hardly have had any value if all these peoples were arbitrarily inferior to the Greeks.

The Reception and Later History of the Geographika

Although Eratosthenes is best remembered today as the originator of the discipline of geography, in antiquity his reputation as a geographer was less certain. The *Geographika* came under extensive criticism within a generation of his death. By the late Hellenistic period, his topographical data, especially in the western Mediterranean, had been superseded and was seen as inadequate, something that was used to criticize the work as a whole. Eratosthenes became famous for calculating the circumference of the earth (a feat not originally published in the *Geographika*), yet he would be remembered primarily as a philologist and poet: the *Souda* biography refers neither to the *Geographika* nor geography, although three other works and four other academic disciplines are cited. Plutarch saw Eratosthenes as a polymath who was useful for a number of topics,[110] not including geography, and placed him alongside Alkaios and Euripides, quoting his poetry.[111] The obsessive bibliophile Athenaios mentioned Eratosthenes 15 times, always as philologist or poet,[112] but showed no knowledge of the *Geographika*: it is difficult to imagine him not citing a title if he knew it.

Evaporation of Eratosthenes' reputation as a geographer may have been assisted by the attacks on the *Geographika* that appeared soon after its publication. The earliest known critic is a certain Polemon,

---

own work on Alexander: see W. W. Tarn, *Alexander the Great* 2: *Sources and Studies* (Cambridge 1948) 437–49. Eratosthenes' youth in Kyrene, one of the most multicultural Greek states, may also have influenced his thought (Fraser, *PA* vol. 1, pp. 530–1).

[110] Plutarch, *Demosthenes* 9, 30; *Alexander* 3, 31; *Themistokles* 27; *Lykourgos* 1; *Lives of the Ten Orators* 8; *On the Fortune and Virtue of Alexander* 8; *Whether an Old Man Should Be in Public Affairs* 3.

[111] Plutarch, *Symposiakon* 7.1.3; *Whether Land or Sea Animals Are More Clever* 32; *On Stoic Discrepancies* 29; *Greek and Roman Parallel Tales* 9.

[112] Athenaios 1.2a, 16d, 24b; 2.36ef, 41d; 4.140a, 154a; 5.189d; 7.276ab, 281c, 284d; 9.376b; 10.418a; 11.482a, 499e.

probably Polemon of Ilion, who was active in the beginning of the second century BC and whose geographical work focused on the monuments of Greece.[113] He may have limited his objections to Eratosthenes' topographical details and his disinterest in fieldwork.[114] Somewhat later a full-fledged critique was made by Hipparchos of Nikaia, who was born about the time of Eratosthenes' death and was active until at least 126 BC.[115] As an astronomer and mathematician Hipparchos was disturbed by inconsistencies in Eratosthenes' calculations, especially his terrestrial measurements. Moreover, Hipparchos believed that mathematics and astronomy, rather than overland distances obtained by hearsay from travelers, should be the basis of geographical scholarship.[116] Hipparchos had mathematical and astronomical skills not available to Eratosthenes that could be used for geographical research, and thus he found his techniques not only erroneous but obsolete. Hipparchos' objections were published under the title *Against the Geography of Eratosthenes*, whose three books mirrored the structure of the very work he was dismissing;[117] the treatise was used extensively by Strabo, in whose own *Geography* 55 of the 63 identified fragments appear. Strabo, whose sympathies were more toward Eratosthenes than Hipparchos, tended to uphold the former against the latter, but not always.

Hipparchos established that there were difficulties in Eratosthenes' methodology, yet some later geographical writers still found him a reliable source. The geographer of the early second century BC Agatharchides of Knidos excerpted Eratosthenes heavily in his own work on the Erythraian Sea and even quoted him directly.[118] The author of the *Periplous Dedicated to King Nikomedes*, from the latter second century BC, was inspired by his great precision, considering him the source that he found most convincing.[119] His contemporary Polybios agreed with Eratosthenes in a number of instances yet found serious fault with his interpretation of Homer and his handling of material about the west and

[113] Plutarch, *Symposiakon* 5.2; on Polemon, see James G. Frazer, *Pausanias's Description of Greece* (reprint, New York 1965) vol. 1, pp. lxxxiii–xc.

[114] Strabo 1.2.2.

[115] For the details of his career and evidence of his date see Dicks, *Hipparchus* 1–10.

[116] See, for example, Strabo 2.1.1–11 (= Hipparchos F12–15); Dicks, *Hipparchus* 32.

[117] Strabo 1.1.12 (= Hipparchos F11), 2.1.41. Hipparchos' title suggests a critique of Eratosthenes' theory of geography, not merely his treatise.

[118] See the direction quotation of Eratosthenes in Agatharchides, F8 (from Eratosthenes, F2).

[119] *Periplous* 112–14. This text is also known, erroneously, as Pseudo-Skymnos: see the edition in *Les Géographes Grecs* (ed. Didier Marcotte, Paris 2002).

northwest of Europe, largely because of Eratosthenes' reliance on Pytheas of Massalia.[120] Polybios' intense dislike of Pytheas caused him to be ill-disposed toward Eratosthenes, yet it must also be said that in Polybios' day, especially after the fall of Carthage in 146 BC, knowledge of the western Mediterranean and northwest Europe far surpassed Eratosthenes' information, and many of his details would have seemed naïve. Similarly, Artemidoros of Ephesos, writing at the very end of the century with personal knowledge of the far west and Atlantic, rejected some of Eratosthenes' data from those regions.[121]

These criticisms of Eratosthenes, especially by Hipparchos and Polybios, caused the fading of his reputation as a geographer by the early first century BC, when a certain Serapion of Antioch was a vigorous critic.[122] At roughly the same time Poseidonios' *On the Ocean* seems to have made little use of Eratosthenes' treatise. A century after Eratosthenes' death, geographical scholarship had moved in other directions.

Yet there seems to have been a copy of the *Geographika* available in Rome in the mid-first century BC, perhaps belonging to Cicero's associate T. Pomponius Atticus.[123] Cicero was planning to model his own geographical work on the treatise, Caesar consulted it, and Varro and Vitruvius also seem to have been aware of it.[124] Vergil, whose scholarly depth is astonishing, made use of it for the *Aeneid*.[125] By the 20s BC Strabo was writing his own *Geography*, which thoroughly critiqued both Eratosthenes' treatise and those of his major critics, Hipparchos and Polybios, thus preserving many of the extant fragments of Eratosthenes' work but often in a manner so tangled between the three that comprehension of his ideas can only be grasped imperfectly. How long the actual text survived after the Augustan period is not certain. There is a steady stream of citations, but only three authors seem to have had any familiarity with the treatise. Pliny the Elder cited the work 16 times (F27–8, 41–2, 70, 76, 93, 97, 101, 103, 105, 111, 115–16, 126–7). Arrian made only three

---

[120] Polybios 34.1.16, 2.11, 5.8, 5.12, 9.4, 15.6. On Polybios and Pytheas, and the reasons behind the animosity, see Roller, *Pillars* 66–7.

[121] Strabo 3.2.11, 17.3.2. On Artemidoros' date, see Markianos of Herakleia, *Epitome of the Periplous of Menippos* 3 (*GGM* vol.1, p. 566).

[122] Cicero, *Letter to Atticus* 2.6 (#26, from April 59 BC).

[123] Cicero, *Letter to Atticus* 2.4, 6 (#24, 26, from April 59 BC) and F. W. Walbank, "The Geography of Polybius," *ClMed* 9 (1947) 155.

[124] Cicero, *Letter to Atticus* 2.6 (#26, April 59 BC); Caesar, *Gallic War* 6.24; Varro, *de re rustica* 1.2.3–4; Vitruvius 1.1.17, 6.9, 11. Vitruvius considered Eratosthenes one of the most learned scholars of antiquity, a type of person rarely encountered. Both Cicero and Vergil made use of Eratosthenes' semi-geographical poem *Hermes* (supra, p. 14).

[125] Infra, p. 14.

known references (F23, 71–2) but praised Eratosthenes' broad vision.[126] Dionysios Periegetes wrote the *Circuit of the Inhabited World*, a hexameter poem of over 1,000 lines, which made extensive use of Eratosthenes' treatise.[127] Arrian and Dionysios, both of the early second century AC, are the last who seem to have consulted the *Geographika* directly. More significant, perhaps, is the failure of the geographical authors Pomponius Mela and Ptolemy, the polymath Plutarch, or the bibliophile Athenaios to have any awareness of it, indicating that even by the first century AC the text had become rare. Strabo's *Geography*, which contained so many citations of the *Geographika*, became equally elusive: unknown to Pliny, Plutarch, Ptolemy, and Pausanias, a single copy, perhaps acquired by Dionysios Periegetes, ended up in Alexandria in the second century AC, but was not widely used until Byzantine times.[128] So even the major source for Eratosthenes' treatise remained hard to come by, and the occasional later references to the *Geographika*, into Byzantine times, are derivative and have no concept of the work as a whole, only using it for narrow specialized purposes, often linguistic. What was remembered best, in addition to his reputation as a poet and philologist, was Eratosthenes' determination of the circumference of the earth, appearing in his treatise *On the Measurement of the Earth*. Like the *Geographika*, this survived into Roman times: the most extensive excerpts were made by the mathematical authors Geminos, probably of the early first century BC, and Kleomedes, of the following century or later.[129] Eratosthenes' measurement (unlike anything from the *Geographika*) was cited steadily and repeated into late antiquity by authors such as Vitruvius, Pliny, Ptolemy, Censorinus, and many others, becoming part of the lore of ancient scholarship.[130] Most of these mathematical authors referred to the *Measurement* in varying degrees of detail, inevitably recording the circumference of 252,000 stadia.

Because from early Roman Imperial times there was no text of the *Geographika* to be preserved, modern editions of its fragments have

[126] Arrian, *Indika* 3.1.

[127] See Christian Jacob, *La description de la terre habitée de Denys d'Alexandrie* (Paris 1990).

[128] Aubrey Diller, *The Textual Tradition of Strabo's Geography* (Amsterdam 1975) 6–11.

[129] Geminos 16.6–9 (for his date, see Evans and Berggren [supra n. 71] 15–22); Kleomedes 1.7, 2.1 (for his date, see *Cleomedes' Lectures on Astronomy* [ed. Alan C. Bowen and Robert B. Todd, Berkeley 2004] 2–4).

[130] Vitruvius 1.6.9, 11; Pliny 2.247; Ptolemy, *Mathematical Syntaxis* 1.67.22; Censorinus, *On the Birthday* 13.2; 15.2.

been late and rare. The first attempt to salvage the treatise from those
who quoted it was by Günther Carl Fridrich Seidel (1764–1800), who
also wrote on the American Revolution, and whose *Eratosthenis geographicorum fragmenta*, his dissertation at Göttingen, was published
there in 1789. Seidel established the basic three-book pattern and identified about 100 fragments, appending astonishingly long commentaries.
He also presented the fragments of the *Measurement* separately, as an
appendix to Book 2. Thirty years later, the first complete edition of the
entire corpus of the fragments of Eratosthenes was completed by Gottfried Bernhardy (1800–1875) as his dissertation for the University of
Berlin (*Eratosthenica*, Berlin 1822). Bernhardy identified 127 fragments
of the *Geographika* and also structured them according to the three
books of the work. The fragments of *On the Measurement of the Earth*
were included without distinction and Bernhardy was unaware of that
work's separate history. Other than these particular fragments, most of
Bernhardy's choices were valid. The attached commentaries are brief.

Most of Bernhardy's fragments were included by Ernst Hugo Berger
(1836–1904) in his 1880 edition of Eratosthenes' geographical material.[131] He created the first thorough and critical edition, identifying
295 fragments and dividing them topically among the three books of the
treatise, with 170 coming from the third book. He also presented 20 testimonia and a summary of critical issues, especially the effect of Hipparchos and Strabo on comprehension of the work. There are no individual commentaries, but detailed summary discussions after groups of
fragments. Although Berger's work is now out of date and can easily be
criticized, his efforts were prodigious in pulling out these many citations from nearly 50 ancient authors. Nevertheless there are some issues that make modern use of his work difficult. Perhaps the largest
problem is the continued failure to recognize *On the Measurement of the
Earth* as a separate treatise, placing all of its fragments in his collation
of Book 2. Moreover, in the fashion of his era, he broke up continuous
discussions into several fragments (often separating them), isolating
points where Eratosthenes was mentioned by name—a fatal practice
insofar as the material from Strabo is concerned—and including many
repetitive citations from late antiquity. He also made some unfortunate
judgments about Eratosthenes' sources, assuming, for example, that
any citation by Strabo of the Seleukid explorer Patrokles must have

---

[131] *Die Geographischen Fragmente des Eratosthenes* (Leipzig 1880). Berger was a
seminal figure in the study of ancient geography, eventually appointed professor of historical geography at Leipzig.

been through Eratosthenes, even if he were not mentioned. On the other hand, Berger validated Strabo's dismissal of the data of Pytheas of Massalia. Yet Berger's choice of fragments is extremely thorough and few need to be added, although astonishingly he missed the vital and detailed discussion of the continents and how they are divided (F33). In addition, the number of Berger's fragments can be reduced by over a hundred (without lessening the amount of material), excising those that are repetitive or belong to the *Measurement*, as well as creating longer continuous passages. His edition and Seidel's are the only ones before the present work.[132] Since Berger's time, efforts have largely been confined to critical points within the *Geographika* itself, mostly, as is inevitable with geographical authors, topographical analysis. Overall consideration of the treatise has been scant, with major exceptions A. Thalamas, P. M. Fraser, and the recent work of Klaus Geus and Germaine Aujac.[133] Since most of the fragments are in the *Geography* of Strabo, commentators on that author are inevitably drawn into scrutiny of Eratosthenes' treatise, but to differing degrees and with their own agenda: as noted elsewhere, understanding Strabo and his use of sources is a complex issue in itself.

Historians of science have their own interest in Eratosthenes, centered on the mathematical and astronomical issues apparent in the *Measurement*.[134] When the *Geographika* is considered, it is in a numerological sense, often to the exclusion of topographic, ethnographic, or historical matters. Although it is perhaps of passing interest to note whether Eratosthenes' calculations conform to modern ones, to focus on this seems not to understand the whole of the work. It is often overlooked that most of the distances come not from astronomical calculations but itineraries and travelers' reports.[135] Eratosthenes showed great skill in the *Measurement*, but the *Geographika* is about neither mathematics or

[132] R. M. Bentham's unpublished Ph.D. thesis, "The Fragments of Eratosthenes," University of London, 1948, is inaccessible to all but a few.

[133] Thalamas (supra n. 103); Fraser, *PA* vol. 1, pp. 525–39; Geus, *Eratosthenes* 260–88; Aujac (supra n. 4). Other important works include Bunbury (supra n. 103) vol. 1, pp. 615–50 (still of great value), and Germaine Aujac, *Strabon et la science de son temps* (Paris 1966) 49–64.

[134] See, for example, Jacques Dutka, "Eratosthenes' Measurement of the Earth Reconsidered," *AHES* 46 (1993) 55–66; Dennis Rawlins, "Eratosthenes' Geodesy Unraveled: Was There a High-Accuracy Hellenistic Astronomy?" *Isis* 73 (1982) 259–65; and his "The Eratosthenes-Strabo Nile Map," *AHES* 26 (1982) 211–19.

[135] See, refreshingly, Bernard R. Goldstein, "Eratosthenes on the 'Measurement' of the Earth," *HM* 11 (1984) 411–16.

astronomy, and those who see in the treatise more than a casual application of these disciplines regrettably have missed its point.

The present work is largely based on Berger's collection of the fragments, with the exclusions and additions noted above. The fragments
have been reordered somewhat (although Berger's judgment here was
generally sound) and renumbered straight through, eliminating his
confusing three-tier system. The English version is the first ever. Although most of the original sources have been translated within their
own texts, these are of varying quality and usefulness and may not be
sensitive to the work of Eratosthenes. For example there have only been
two complete English editions of Strabo, that by H. C. Hamilton and
W. Falconer (Bohn's Classical Library, 1854–7), which is too early to be
of great value today, and the Loeb, published between 1917 and 1932
but started by J. R. S. Sterrett before 1914 (who was responsible for
Books 1 and 2, where most of the citations of Eratosthenes occur) and
completed by H. L. Jones. This edition is not only nearly a century old
but is heavily flawed both in the Greek text and the translation.

Moreover, the "fragments" are rarely direct quotations of Eratosthenes, or even paraphrases, but synthetic arguments that bring together
material from several treatises.[136] Especially in the case of the information preserved by Strabo, it is not always possible to identify the particular source. One must make a careful path between too narrow a choice
and too broad. Mention of Eratosthenes by name has always been a
valuable criterion but it is not an absolute one, especially in the case of
Strabo's many verbs without subjects. In the current edition an attempt
has been made to choose a middle path, with the understanding that
the possibility of error is in both directions: there is the certainty that
some material originating with Eratosthenes is not included, as well as
data not from Eratosthenes remaining within the chosen fragments.
Because Strabo was blending Eratosthenes' treatise into his own agenda,
his handling of the text was far from linear. Issues and concepts introduced by Strabo may only be explained later. Points may be repeated.
Strabo, writing 200 years after Eratosthenes, could be insensitive to the
novelty of many of his ideas. Eratosthenes wrote at a time when the
Roman presence in the Greek world was just beginning to make itself
felt (the *Geographika* was probably written between the First and Second Punic Wars). On the other hand, Strabo lived in an era where long
years of Roman expansionism had not only affected the political dynam-
ics of the Mediterranean (and thus its geography) but resulted in all of

---

[136] See, for example, Strabo 1.2.1–14.

its coast and much of western Europe becoming understood rather than remaining remote and exotic. Since it can be impossible to separate out the actual thoughts of Eratosthenes from Strabo's often lengthy reanalyses, the fragments can seem repetitive. Their ordering may not be correctly presented, although the general outline of Eratosthenes' treatise is clear. The summaries that precede each commentary are designed to pull Eratosthenes' thoughts out of such tangles. There is the hope that all the fragments of the *Geographika* that can be identified have been, and the problems are discussed in detail in the individual commentaries, which can separate out the contributions of Eratosthenes in a way not immediately apparent in the continuous translations of fragments.

Most of the translation difficulties concern the numerous fragments from the *Geography* of Strabo. His highly elliptical style causes problems for the interpreter and translator. In addition to verbs without obvious subjects, subordinate clauses may be reduced to the minimum, often using a pronoun-adjective combination standing alone without nouns or verbs. Meaning is often unclear despite the inflections of the Greek language. Less serious are the frequent omission of the word "stadia" and the inconsistent spelling of toponyms: a modern tendency to regularize such spellings only adds to the confusion, as such differences are revealing about Strabo's immediate source. The style is varied, to say the least, sometimes highly elliptical to the point of obscurity, and elsewhere, perhaps even in the same passage, excessively repetitive, using the same verb or noun several times in a few lines. This translation has attempted to produce the vagaries of Strabo's style as accurately as possible, without any attempt at spurious elegance. Bracketed words are inserted when necessary for clarity, especially in regard to the subjects of verbs, but this has been kept to the minimum, although such may not seem to be the case. The few Latin fragments (mostly from Pliny the Elder) have the additional problem of the latinizing of toponyms that may be obscure in any language and exist only in oblique cases, so restoration of the nominative may be hypothetical. Pliny was not always consistent in his latinization, again perhaps reflecting his source of the moment, and he often retained Greek case endings. When any of these issues becomes a detriment to comprehension it is discussed in the commentary.

ERATOSTHENES

Geographika

Book 1

Introduction

1 (IA1). Strabo, Geography *1.1.1.*
That which we choose to investigate now, geography, is, we believe, a discipline like others and for the scholar. We believe that this is not inconsequential and that it is obvious for many reasons. Those who first dared to begin to consider it were men such as Homer, Anaximandros of Miletos and his fellow-citizen Hekataios, just as Eratosthenes has said, as well as Demokritos, Eudoxos, Dikaiarchos, Ephoros, and a number of others.

Homer and Geography

2 (IA4, IA19, IA21). Strabo, Geography *1.2.3.*
He [Eratosthenes] says that all poets attempt to amuse rather than teach. On the contrary, the ancients say that poetry is foremost a pursuit of knowledge, introduced into our life from youth, which teaches us with pleasure about character, emotion, and actions. Moreover, we say that only the poet is wise. For this reason the Hellenic cities educate their youth first of all in poetry, not presumably for the sake of mere amusement but to learn morality. Even musicians, teaching plucking, lyre playing, and flute playing, claim this virtue, for they say that such an education improves character. One may hear this said not only by the Pythagoreans, but Aristoxenos maintains the same thing. And Homer said that the singers were chastisers, as in the case of the guardian of Klytaimnestra,

> whom Atreides, going to Troy, strictly commanded to guard
> his wife
>
> *[Odyssey 3.267–8],*

and Aigisthos was not able to prevail over her before

> he took the singer to a deserted island and left him there,
> and then willingly led her, willing, to his home
>> *[Odyssey 3.270–1].*

Apart from this, Eratosthenes contradicts himself. Shortly before he said this, at the beginning of his geographical treatise, he says that from earliest times all of them [the poets] have eagerly placed themselves in the mainstream of that discipline. For example, [Eratosthenes says] whatever Homer learned about the Aithiopes he recorded in his poem, as well as about the Egyptians and Libyans. Insofar as Hellas and the neighboring places are concerned, he [says that he] elaborated excessively, saying that Thisbe is abounding in doves [*Iliad* 2.502], Haliartos is grassy [503], Anthedon is the remotest [508], Lilaia is by the Kephissian springs [523], but also [says] that he never threw out a useless qualification. Is someone who does this an entertainer or a teacher? By Zeus the latter, you say, but that which is beyond perception he [Homer] and others have filled with legendary marvels. He [Eratosthenes] should have said that every poet writes for the sake of mere entertainment and teaching, but he said "merely for entertainment and not for teaching." He meddles still further when he asks how it contributes to the quality of the poet to become skilled in places or military command or farming or rhetoric or whatever else others might wish him to have acquired. The desire for him to acquire everything would be going beyond the proper limit in ambition, just as if someone, as Hipparchos says, were to hang apples and pears on Attic wreaths, which cannot hold them, burdening him with all knowledge and every skill. You may be right, Eratosthenes, about that, but you are not right when you take away from him [Homer] his great learning, and declare that his creativity is the mythology of an old woman, who has been allowed to fabricate (as you say) for her own amusement whatever appears.

3 (IA11). Strabo, Geography *1.2.7.*
But he [Homer] does not only speak of nearby places—as Eratosthenes says—those within Hellas, but also many far away. Homer tells myths more accurately than later mythological authors, not totally recounting marvels, but for the sake of knowledge, using allegory, revision, and popularity, especially concerning the wanderings of Odysseus, about which he [Eratosthenes] makes many mistakes, maintaining that the commentators and the poet himself are nonsense.

4 (IA17). Strabo, Geography *1.2.17.*
To fabricate everything is not plausible, and not Homeric. Everyone believes that his poetry is a scholarly treatise, not like Eratosthenes says,
who commands us neither to judge the poems in regard to their thought,
nor to seek history in them.

5 (IA16). Strabo, Geography *1.2.15.*
He [Polybios] does not approve of this assertion by Eratosthenes, where
he says that one will find where Odysseus wandered when you find the
cobbler who sewed up the hide of winds.

6 (IA2, IA12, IB3, IIIB115). Strabo, Geography *1.2.11–4.*
(11) Having set forth these preliminaries, it is necessary to ask what is
meant by those who say that Homer placed the wanderings of Odysseus
around Sikelia and Italy. This can be understood in two ways, one better
and the other worse. The better is to accept that he believed that the
wanderings of Odysseus were there, and taking this as the truth, elaborated this assumption poetically. One could naturally say this about
him, and one would find vestiges of the wanderings—and those of many
others—not only around Italy but in the farthest regions of Iberia. The
worse interpretation is to accept the elaboration as historical, because
Okeanos, Hades, the cattle of Helios, hospitality by goddesses, transfor-
mations, large Kyklopes and Laistrygonians, the shape of Skylla, the
distances sailed, and many other similar things are clearly writings
about marvels. But it is not worth refuting someone [Eratosthenes] who
is so clearly in error about the poet, as if one could not say that the return of Odysseus to Ithaka and the Slaughter of the Suitors and the
battle in the country between them and the Ithakans happened in that
very way, and it is also not proper to attack someone who interprets it
literally.
   (12) Eratosthenes has confronted both of these reasons, but not well.
In the second case, he believes that he [Homer] attempts to misrepre-
sent something obviously false and unworthy of a lengthy discussion,
and in the former, that all poets tell falsehoods and that their experience of places or the arts does not lead to virtue. Myths are related
about uninvented places, such as Ilion, Ida, and Pelion, but also about
invented ones, such as where the Gorgons and Geryon are. He [Eratos-
thenes] says that those mentioned in the wanderings of Odysseus are
also a construct, and that those who say they are not invented but sub-
stantiated are convicted of falsehood because they do not agree with one
other. At any rate, some put the Seirenes on Pelorias, and others on

the Seirenoussai, more than 2,000 stadia away, allegedly the three-headed promontory that separates the Kymaian and Poseidonian Gulfs. But this rock is not three-pointed, nor does its summit come at all to a head, but a kind of elbow projects out, long and narrow, from the district of Syrrenton to the strait of Kapria, having the sanctuary of the Seirenes on one side of the mountainous ridge, and on the other (toward the Poseidonian Gulf) lie three little deserted and rocky islands that are called the Seirenai. At the strait itself, where the elbow is narrow, is the sanctuary of Athene.

(13) Moreover, even if those who have handed down the account of those places are not in agreement, frankly we should not throw out the entire account, since on the whole it may be more believable. As an example, I would ask whether it is said that the wanderings were around Sikelia or Italy, and that the Seirenes are somewhere around there. The one who says that they are in Pelorias disagrees with the one putting them on the Seirenoussai, but both of them do not disagree with someone saying that they are around Sikelia and Italy, but give him greater credibility, for although they do not point out the same place, nonetheless they do not depart from the region of Italy and Sikelia. If, then, someone were to add that a memorial of Parthenope, one of the Seirenes, is shown in Neapolis, there is even more credibility, although this is mentioning a third place. Moreover, Neapolis lies in this gulf, which Eratosthenes calls the Kymaian and which is formed by the Seirenoussai, and thus we can believe more strongly that the Seirenoi were around these places. The poet did not learn about each accurately, nor do we seek accuracy from him, but nonetheless we do not assume that he learned to sing about the wanderings without [knowing] where or how they happened.

(14) Eratosthenes infers that Hesiod learned that the wanderings of Odysseus were throughout Sikelia and Italy, and believing this, recorded not only those places mentioned by Homer, but also Aitna, Ortygia (the little island next to Syrakousai), and Tyrrhenia, and that Homer did not know them and did not wish to put the wanderings in known places. Are Aitna and Tyrrhenia well known, but Skylla and Charybdis, Kirkaion, and the Seirenoussai not at all? Is it fitting for Hesiod not to talk non-sense and to follow prevailing opinions, yet for Homer "to shout forth everything that comes to his untimely tongue"? Apart from what has been said concerning the type of myth that it was fitting for Homer to relate, most of the prose authors who repeat the same things, as well as the customary local reports about these places, can teach that these are not fantasies of poets or even prose authors but vestiges of peoples and events.

7 (IA14). Strabo, Geography *1.2.18–19.*

(18) When did a poet or prose author persuade the Neapolitans to make a memorial for Parthenope the Seiren, or those in Kyme, Dikaiarchia, and at Baiai for Pyriphlegethon, the Acherousian Marsh, the oracle of the dead at Aornos, or Baios and Misenos, both companions of Odysseus? It is the same with the Seirenoussai, Porthmos, Skylla and Charybdis, and Aiolos. This must not be scrutinized carefully or considered without roots or a home, not attached to the truth or any historical benefit.

(19) Eratosthenes himself suspected this, for he says that one might understand that the poet wished to put the wanderings of Odysseus toward western places, but set aside the idea, because he had not learned about them accurately, or because he chose not to do so, in order to develop each element more cleverly and more marvelously. He [Eratosthenes] understands this correctly, but is wrong in regard to why he did it, since it was not for silliness but for a benefit. Therefore it is proper that he [Eratosthenes] should undergo examination about both this and why he says that marvelous tales are told about faraway places because it is safe to tell falsehoods about them.

8 (IA5, IA6, IB4). Strabo, Geography *7.3.6–7.*

(6) What Apollodoros says in the preface to the second book of his *On Ships* is not acceptable. He approves of what Eratosthenes asserted, that both Homer and other ancient authors knew Hellenic places, but they were ignorant of those far away, ignorant of long journeys, and ignorant of sea voyages. In support of this he [Apollodoros] says that Homer calls Aulis rocky [*Iliad* 2.496], just as it is, Eteonos many-ridged [497], Thisbe abounding in doves [502], and Haliartos grassy [503], but that neither he nor the others knew faraway places. There are about 40 rivers that flow into the Pontos, but he does not mention even those that are the best known, such as the Istros, Tanais, Borysthenes, Hypanis, Phasis, Thermodon, or Halys. Moreover, he does not mention the Skythians, but creates certain "noble Hippemolgoi" or the "Galaktophagoi" and the "Abioi." Concerning the Paphlagonians of the interior, his report is from those who approached these territories on foot, but he is ignorant of the coast, and naturally so. At that time the sea was not navigable and was called the Axinos because of its wintriness and the wildness of the peoples living around it, most of all the Skythians, who sacrificed strangers, ate their flesh, and used their skulls as drinking cups. Later it was called the Euxeinos, when the Ionians founded cities on its coast. Moreover, he is ignorant of matters concerning the Egyptians and

Libyans, such as the rising of the Nile and the silting up of the sea (which he records nowhere), or the isthmus between the Erythraian and Egyptian seas, or Arabia, Aithiopia, and the Ocean, unless one should agree with the scholar Zenon when he wrote:

> I came to the Aithiopes and Sidonians and Arabians
>   [Homer, Odyssey 4.84, emended].

But this is not surprising for Homer, for those more recent than he have been ignorant of many things and tell of marvels. Hesiod speaks of the Hemikynes, the Megalokephaloi, and the Pygmaioi; Alkman about the Steganopodes; and Aischylos about the Kynokephaloi, the Sternophthalmoi; and the Monommatoi—he says this in the *Prometheus*—and countless others. From these he [Apollodoros] proceeds to the writers who speak of Mount Rhipaia and Mount Ogyion, and the settlements of the Gorgons and Hesperides, the Meropian land of Theopompos, the Kimmerian city of Hekataios, the Panchaian land of Euhemeros, and Aristotle's river stones formed from sand but melted by rain. In Libya there is the city of Dionysos that no one can find twice. He censures those who say that the wanderings of Odysseus were, according to Homer, around Sikelia, for if so, one must say that although the wanderings were there, for mythological reasons they were placed by the poet in the Ocean. Although others can be excused, Kallimachos cannot be at all, in his pretense as a scholar, who says that Gaudos is the Island of Kalypso and Korkyra is Scheria. He [Apollodoros] accuses others of being mistaken about Gerena, Akakesion, Demos in Ithaka, Pelethronion in Pelion, and Glaukopion in Athens. To these he adds some minor things and then ceases, having transferred most of them from Eratosthenes, which, as I have said previously, are not correct. In regard to Eratosthenes and him [Apollodoros], one must grant that more recent writers are more knowledgeable than the ancient ones, but thus to go beyond moderation, particularly in regard to Homer, seems to me something for which they could justly be rebuked, and indeed one could say the opposite, that when they are ignorant themselves about these things, they make reproaches at the poet. What remains on this topic happens to be mentioned at the appropriate places as well as generally.

(7) I was speaking now about the Thracians:

> The Mysoi fighting hand-to-hand and the illustrious
> Hippemolgoi, Galaktophagoi, and Abioi, the most just of men
>   [Homer, Iliad 13.5–6],

wishing to compare what was said by myself and Poseidonios with them [Eratosthenes and Apollodoros]. In the first case, the reasoning that they made is opposite to what they proposed. They propose to demonstrate that those earlier were more ignorant of places far from Hellas than were those more recent, but they showed the opposite, not only about far places, but also those within Hellas. But, as I was saying, let us postpone the rest and observe that which is here: they say that because of ignorance he does not mention the Skythians or their cruelty toward strangers, whom they sacrifice and eat their flesh, using their skulls for drinking cups, and because of whom the Pontos was called the Axenos, but he creates certain "illustrious Hippemolgoi, Galaktophagoi, and Abioi, the most just of men," who are nowhere on earth. How could there be the name "Axenos" if they did not know about their savageness, and that they were the most [savage] of all? These are presumably the Skythians. Were not the Hippemolgoi beyond the Mysoi and Thracians and Getai, as well as the Galaktophagoi and Abioi? Even now there are Amaxoikoi and Nomades, as they are called, who live off their animals and milk and cheese, especially that from horses, not knowing about storing things or trading, except goods for goods. How could the poet be ignorant of the Skythians if he spoke of certain Hippemolgoi and Galaktophagoi? At that time they were called the Hippemolgoi, and Hesiod is a witness to this, in the words that Eratosthenes quotes:

> Aithiopes, Ligyes, and also the mare-milking Skythians
> *[Catalogue, F40].*

9 (IA3). Strabo, Geography *1.2.37.*
Apollodoros, in censuring Kallimachos, agrees with those around Eratosthenes, because, although a scholar, he named Gaudos and Korkyra, in opposition to the Homeric assumption that places were located in the External Ocean, where he says that the wanderings were.

10 (IA7, IA8, IB1). Strabo, Geography *1.2.22–4.*
(22) Continuing in his assumption about the falsity of Homer, he [Eratosthenes] says that he does not even know that there are several mouths to the Nile, or its name, although Hesiod knows, for he records it. Concerning the name, it is probable that it was not yet used in his time. In regard to the mouths, if they were unnoticed or only a few knew that there were several rather than one, one might grant that he did not know this.

[Further elaboration, and Homeric knowledge of remote places, omitted.]

(23) How they [Eratosthenes and others] reproach him [Homer] about this island of Pharos is unreasonable, because he says it is in the open sea, as if he were speaking from ignorance.

[Refutation of this statement omitted.]

(24) The same mistake is made concerning his [Homer's] ignorance of the isthmus between the Egyptian sea and the Arabian Gulf, suggesting that he is wrong in speaking of

> The Aithiopes, divided in two, the farthest of men
> > *[Odyssey 1.23]*.

To speak of this is correct, and later writers do not rebuke him justly.

11 (IA10). Strabo, Geography *1.2.20–1.*
The poet spoke accurately:

> Boreas and Zephyros, the Thracian winds
> > *[Iliad 9.5]*.

But he [Eratosthenes] does not accept this correctly and quibbles about it, as if he were speaking generally that the Zephyros blows from Thrace, yet he is not speaking generally, but about when they come together on the Thracian sea around the Gulf of Melas, which is a part of the Aegean itself. For Thrace, where it touches Makedonia, takes a turn to the south, and forms a promontory into the open sea, and it seems to those on Thasos, Lemnos, Imbros, Samothrake, and the surrounding sea that the Zephyroi blow from there, just as for Attika they come from the Skeironian rocks, because of which the Zephyroi, and especially the Argestai, are called the "Skeirones." Eratosthenes did not perceive this, although he suspected it. Nevertheless he told about the turn of the land that I have mentioned. He accepts it [what Homer said] as universal and then accuses the poet of ignorance, that the Zephyros blows from the west and Iberia, but Thrace does not extend that far. Is he [Homer] really unaware that the Zephyros blows from the west?

[Discussion of winds omitted.]

These are the corrections that are to be made at the beginning of the first book of the *Geographika*.

The History of Geography

12 (IB5). Strabo, Geography *1.1.11.*
Let what has now been said be sufficient, that Homer was the beginning
of geography. It is obvious that his successors were also notable men
and familiar with learning. Eratosthenes says that the first two after
Homer were Anaximandros, a pupil and fellow-citizen of Thales, and
Hekataios of Miletos, and that the former was the first to produce a geographical plan, and Hekataios left behind a treatise, believed to be his
because of its similarity to his other writings.

13 (IB6, IB8, IIA9, IIIB93, IIIB114). Strabo, Geography *1.3.1–2.*
(1) Eratosthenes does not handle the following well: he discusses men
not worthy of remembering, sometimes refuting them, and other times
believing in them and using them as authorities, such as Damastes and
others like him. Even if there is some truth in what they say, we should
not use them as authorities or believe them. On the contrary, we should
use only reputable men in this way, those who have generally been correct, and even if they have omitted many things, or not discussed them
sufficiently, they have said nothing untrue. But to use Damastes as an
authority is no different from invoking as an authority the Bergaian, or
the Messenian Euhemeros and the others that he [Eratosthenes] quotes
in order to discredit their nonsense. He tells one of his [Damastes']
pieces of trash, that he believes that the Arabian Gulf is a lake, and that
Diotimos the son of Strombichos, leading an Athenian embassy, sailed
up the Kydnos from Kilikia to the Choaspes River, which flows by Sousa,
arriving at Sousa on the fortieth day. He was told this by Diotimos himself. Then he wonders how it was possible for the Kydnos to cut across
the Euphrates and Tigris and empty into the Choaspes.

   (2) Not only could one disapprove of this, but additionally because he
[Eratosthenes] says that the seas were not yet known even in his own
time, exhorting us not easily to believe people by chance, rendering at
length the reasons that no one should be believed who tells mythic tales
about the Pontos and Adrias, yet he himself believes people by chance.
Therefore he believed that the Issic Gulf is the most easterly limit of our
sea, but a point at Dioskourias in the extreme recess of the Pontos is
farther east by about 3,000 stadia, even from the measurement of stadia
that he records. In discussing the northern and extreme areas of the
Adrias, he does not abstain from the fabulous. He also believes many
stories about what is beyond the Pillars of Herakles, naming an island

called Kerne and other places that are nowhere shown today, concerning which I will discuss later. He says that the earliest voyages were for piracy or commerce, and not in the open sea, but along the land, like Jason, who abandoned his ships and made an expedition from Kolchis as far as Armenia and Media. Later he says that in antiquity no one dared to sail on the Euxine, or along Libya, Syria, or Kilikia. Now if he says "in antiquity" about those for whom in our time there is no record, I am not about to speak of them and whether they sailed or not. But if he means those who have been recorded, one would not hesitate to say that those in antiquity are shown to have made longer journeys (whether completed by land or sea) than those later, if we pay attention to what has been said.

14 (IB7, IIIB1, IIIB96). Strabo, Geography *2.4.1–2.*
(1) Polybios says that in his European chorography he omits those from antiquity in favor of those who refute them, especially scrutinizing Dikaiarchos and Eratosthenes, who has produced the ultimate work on geography, and Pytheas, by whom many have been deceived, as he asserted that he traveled over the whole of Brettanike that was accessible, reporting that the circumference of the island was more than 40,000 [stadia], and also recording matters about Thoule and those places where there was no longer any land in existence—and neither sea nor air—but something compounded from these, resembling a sea lung in which, he says, the earth, sea, and everything are suspended, as if it were a bonding for everything, accessible neither by foot or ship. He himself saw the lung but tells the rest from hearsay. This is the report of Pytheas, and he adds that when he returned from there he went along the entire coast of Europe from Gadeira to the Tanais.

(2) Now Polybios says that this is unbelievable: how could someone who was a private individual and poor have gone such distances by ship and foot? Eratosthenes was at a loss whether to believe these things, but nevertheless believed him about Brettanike and the regions of Gadeira and Iberia. But he [Polybios?] says that it is far better to believe [Euhemeros] the Messenian than him, for, he says that he sailed only to one country, Panchaia, but he [Pytheas] closely observed the entire north of Europe as far as the boundary of the world, a report no one would believe even if from Hermes. Eratosthenes called Euhemeros a Bergaian but believes Pytheas, even though Dikaiarchos did not believe him. "Dikaiarchos did not believe him" is an absurd statement, as if it were fitting for him [Eratosthenes] to use as a standard the one against whom he has made so many refutations. I have said that Eratosthenes was ignorant of the western and northern parts of Europe. But there

must be leniency toward him and Dikaiarchos, as they had not seen those places, but who would be lenient toward Polybios or Poseidonios? For it is Polybios who calls what they [Eratosthenes and Dikaiarchos] report about the distances in those regions and other places popular judgments, although he is not free from this when he refutes them.

The Formation of the Earth

15 (IB11, IB12, IB13, IB14, IB15). Strabo, Geography 1.3.3–4.
(3) He [Eratosthenes] himself spoke of the great advance made in the knowledge of the inhabited world by those after Alexander and those in his own time, and then proceeded to a discussion about its shape, not of the inhabited world—which would have been more appropriate to his topic—but of the entire earth. That must also be considered, but not out of its place. He says, then, that in its entirely it is spherical, not as if turned on a lathe, but having certain irregularities, and then he lists the numerous changes in its shape that occur because of water, fire, earthquakes, eruptions, and other such phenomena, but he does not preserve the arrangement here. The spherical shape of the entire earth results from the state of the whole, but the changes in form do not change the earth as a whole (such small things disappear in great things), although they create in the inhabited world differences from one time to another, with one and another cause.

(4) He says that this presents a particular issue, for why, two or three thousand stadia from the sea and in the interior, can one see in many places mussel, oyster, and scallop shells, as well as many lagoons, such as, he says, those around the temple of Ammon and along the 3,000 stadia of the road to it? A large quantity of oyster shells and much salt is still found there today, and eruptions of salt water spring up to some height. In addition, pieces of wreckage from seagoing ships are shown, which they say have been thrown out of a chasm, and there are small columns with dolphins dedicated on them, having the inscription "Of the Kyrenaian envoys." Then he says that he praises the opinion of Straton, the scientist, and also that of Xanthos the Lydian. Xanthos says that in the time of Artaxerxes there was so great a drought that the rivers, lakes, and cisterns became dry, and that he had often seen, far from the sea, in Armenia, Matiene, and Lower Phrygia, stones like mollusk shells, sherds like combs [?], the outlines of scallop shells, and a salt lagoon, and because of this he believed that the plains were once the sea. Straton engages further in the matter of causes, for he says that he believes that the Euxine once did not have its mouth below Byzantion,

but the rivers that empty into it forced it and opened it up, so that the water came out the Propontis and the Hellespont. The same thing occurred in our sea, for here the strait beyond the Pillars was broken through when the sea had been filled by the rivers and the former shallows were uncovered by this flooding. He suggests a cause: first, that the external and internal seas have a different water [depth?], and that even today there is a certain undersea ridge running from Europe to Libya, which shows that formerly the interior and exterior could not have been the same. Those around Pontos are especially shallow, but the Cretan, Sikelian, and Sardoan seas are very deep, since the rivers flowing from the north and east are numerous and large, and fill [the sea] with sediment, but the others remain deep. This is why the Pontos sea is the sweetest and why it flows out toward the place that its bed slopes. He believes that the entire Pontos will fill up in the future, if such an influx continues. And even now the area on the left side of the Pontos is already covered with shallow water, such as at Salmydessos and the place called by sailors the Stethes, around the Istros, and the Skythian desert. Perhaps the temple of Ammon was once on the sea but is now in the interior because there has been an outflow of the sea. He suggests that the oracle became so famous and well known with good reason because it was on the sea, but since it is now so far removed from the sea, there is no good reason for its fame and reputation. In antiquity Egypt was covered by the sea as far as the marshes around Pelousion, Mount Kasion, and Lake Sirbonis. Even today when the salty lands in Egypt are excavated, the holes are found to contain sand and mussel shells, as though the land had been submerged and all the territory around Kasion and the place called Gerrha had been covered with shallow water, so that it connected to the Erythraian Gulf. When the sea gave way, they were revealed, although Lake Sirbonis remained, but then it broke through so that there was a marsh. In the same way the shores of what is called Lake Moiris resemble more the shores of the sea than the shores of rivers. One would admit that the greater part of the continents were once flooded at certain times and then uncovered again, and similarly the entire surface of the earth that is now under water is uneven, just as, by Zeus, that which is above water, on which we live, receives all the changes of which Eratosthenes himself speaks. Thus one cannot accuse Xanthos of saying anything unnatural.

16 (1B16, IB19, IB20, IIA8). Strabo, Geography 1.3.11–15.
(11) But he [Eratosthenes] is so ingenuous that even though a mathematician he will not confirm the opinion of Archimedes, who says in his

On Floating Bodies that all calm and quiet water appears to have a spherical surface, with the sphere having the same center as the earth. Everyone who has ever understood mathematics accepts this point of view. He [Eratosthenes] says that the Internal Sea is a single sea, but he does not believe that it has ever been constituted as a single surface, even in neighboring places. He uses engineers as witnesses for this ignorance, although the mathematicians proclaim that engineering is a part of mathematics. He also says that Demetrios attempted to cut through the Peloponnesian Isthmos to supply a passage for his forces to sail through, but was prevented by the engineers who measured carefully and reported that the sea level of the Korinthian Gulf was higher than at Kenchreai, so that if he were to cut through the intervening land, the entire strait around Aigina as well as Aigina itself and the nearby islands would be submerged and the sailing passage would not be useful. This is why narrow straits have strong currents, especially the narrows of Sikelia which, he says, are similar to the high and low tides of the ocean, with the flow changing twice each day and night, and like the ocean there are two floodings and two withdrawals. He says that similar to the flood tide is the current that goes down from the Tyrrhenian to the Sikelian Sea, as though from a higher level, which is called "the descent," and he says that it begins and ends at the same time as the high tides. It also begins around the rising and setting of the moon and it ceases when it reaches either meridian, that above the earth or below the earth. Like the ebb tide is the opposite current—called "the ascent"—which begins when the moon is at either meridian, just like the ebb tide, and ceases when the moon reaches the points of setting and rising.

(12) The flooding and ebbing of the tides has been sufficiently discussed by Poseidonios and Athenodoros, but concerning the rushing back of straits, which is a more scientific discussion than appropriate in this treatise, it is sufficient to say that there is no single explanation for the currents in straits that corresponds to their form, for it would not be that the Sikelian changes twice a day, as he [Eratosthenes] says, and the Chalkidean seven times, or that the one at Byzantion has no change but continues only having an outflow from the Pontic Sea into the Propontis, as Hipparchos reports, and at times is stopped. If there were a single explanation, the reason would not be what Eratosthenes says, that each sea has a different surface. This would also not be the case with rivers unless they have cataracts, but having them, they do not flow back but go continuously lower. This happens also because the stream and its surface are inclined. But who would say that the surface of the sea is inclined? This is especially because of the theory that the

four bodies—which we would call elements—are made spherical. Thus it does not flow back but also does not become calm and remain so, since they flow together, without a single surface, but one at a higher level and the other at a lower one. It is not like the earth, whose state has assumed a solid form, thus having permanent hollows and protuberances, but water, through the effect of its weight, is carried upon the earth, having the kind of surface that Archimedes says.

(13) He [Eratosthenes], adding to what he has said about Ammon and Egypt, believes that Mount Kasion was once washed by the sea and that the entire region, where what is now called Gerrha is, was covered with shallow water since it was connected with the Erythraian Gulf, becoming uncovered when the seas came together. To say that the place was covered with shallows and connected to the Erythraian Gulf is ambiguous, since "connected" means "to be near" or "to touch," so that, if it is a body of water, one flows into another. I believe that the shallows came near to the Erythraian Sea while the narrows at the point of the Pillars were closed, and the withdrawal happened because of the lowering of our sea due to the outflow at the Pillars. But Hipparchos argues that "connected" is the same as our sea "flowing into" the Erythraian, because of filling up. He demands to know why, when, because of the outflow at the Pillars and our sea changing direction, the Erythraian, which was flowing into it, remained at the same level and was not lowered? According to Eratosthenes himself the entire External Sea flows together and thus the Western and Erythraian Sea are one. Saying this, he insists that the sea outside the Pillars, the Erythraian, and even that which flowed together into it, have the same height.

(14) But Eratosthenes says that he has not recorded that the flowing together with the Erythraian happened at the time of the filling, but only that they were near to one another, and it does not follow that one sea that was kept together would have the same height and same surface, as this is not the case in ours, by Zeus, at Lechaion and around Kenchreai. Hipparchos indicates this in his treatise against him. Knowing that this is his opinion, let me speak on his own account against him and let him not presume that someone who says that the exterior sea is one would agree that its level is one.

(15) Saying that the inscription of the Kyrenaian envoys on the dolphin is false, he [Hipparchos] provides a reason that is implausible: that although the founding of Kyrene was in recorded times, no one recorded that the oracle was ever on the sea. Even if no one reported this, we can infer from the evidence that the place was once on the coast, since the dolphins were erected and inscribed by the Kyrenaian envoys. He con-

cedes that with the raising of the seabed the sea flooded as far as the location of the oracle, somewhat more than 3,000 stadia from the sea, but he does not concede that the raised level covered all of Pharos and most of Egypt, as if such a height were not sufficient to cover them also. And, saying that if our sea were filled to such a level before the outbreak at the Pillars happened, as Eratosthenes said, all of Libya and most of Europe and Asia must first have been covered, and he then adds that the Pontos would have begun to flow together with the Adrias in certain places, since the Istros divides in the region of the Pontos and flows into each sea, because of the lie of the land. But the Istros does not have its source in the Pontos district—but, on the contrary, in the mountains above the Adrias—nor does it flow into each sea, but only into the Pontos, and only branching around its mouths. In common with some of those before him he fails to understand his ignorance, as they understood that a certain river with the same name as the Istros broke away and emptied into the Adrias, from whom the people called Istrians, through whose [region] it flows, took their name, and by this way Jason made his return voyage from Kolchis.

17 (IB18). Strabo, Geography *1.2.31.*
But the isthmus [between the Mediterranean and Erythraian Sea] was also not navigable, and what Eratosthenes suggests is not correct. He believes that the breakout at the Pillars had not yet happened [at the time of the Trojan War] and that the Internal Sea joined the External Sea, and since it was higher, covered the isthmus, but when the breakout occurred it was lowered and thus uncovered the land around Kasion and Pelousion as far as the Erythraian Sea.

18 (IB17). Strabo, Geography *16.2.44.*
Eratosthenes says the contrary, that the territory formed a lake [the Dead Sea], most of which was uncovered by an outbreak, as in the sea ["Sea" emended to "Thessaly"; see commentary].

Geographical Fabrications

19 (IB22). Strabo, Geography *1.3.23.*
Next he [Eratosthenes] discusses those who clearly speak of fabricated and impossible things, some of which are in the form of myths and others in the form of history, concerning whom it is not worthy to mention. Yet in this subject matter he should not have considered those who are nonsense. This, then, is his discussion in the first book of his commentaries.

20 (IB21). Strabo, Geography *1.3.22.*
In regard to what Herodotos [4.36] said, that there are no Hyperboreans
because there are no Hypernotians, Eratosthenes says that this argument is ludicrous and would be like the following sophistry: if one were
to say that there are none who rejoice at the misfortunes of others because there are none who rejoice at the good fortune of others. Moreover
it so happens that there are Hypernotians, and anyway Notos does not
blow in Aithiopia but farther down.

21 (IB23). Strabo, Geography *15.1.7.*
Regarding the tales about Herakles and Dionysos, Megasthenes and a
few others consider them trustworthy, but most, including Eratosthenes, find them not to be trusted and legendary, like the tales among the
Hellenes.

22 (IB23). Strabo, Geography *2.1.9.*
Particularly worthy of disbelief are Deimachos and Megasthenes, for
they write about the Enotokoitai and the Astomoi and the Arrinoi, as
well as the Monophthalmoi, Makroskeles, and Opisthodaktyloi. They
have also revived the Homeric tale about the battle between the cranes
and pygmies, who, they said, were three *spithamai* tall. There are also
the gold-mining ants and Pans with wedged-shaped heads and snakes
that swallow both cattle and deer with their horns. Concerning these
things each refutes the other, as Eratosthenes says. For they were sent
to Palimbothra—Megasthenes to Sandrakottos and Deimachos to his
son Amitrochates—as ambassadors, and left such writings as remind-
ers of their travels, persuaded to do so for whatever reason. Patrokles
was not such a person at all, and the other witnesses Eratosthenes used
are not unreliable.

23 (IB24). Arrian, Anabasis *5.3.1–4.*
(1) But I do not completely agree with Eratosthenes of Kyrene, who says
that everything the Makedonians attributed to divine influence was excessively enhanced in order to please Alexander. (2) He says that they
saw a cavern in the territory of the Parapamisadai and heard an indigenous tale about it, or put together their own story, saying that it was
the very cave where Prometheus had been bound, which the eagle fre-
quented to feed on Prometheus' innards, and that Herakles, coming to
the same place, killed the eagle and released Prometheus from his bonds.
(3) In their version, the Makedonians transferred Mount Kaukasos from
the Pontos toward the eastern part of the world, and the Parapamisadai

territory as far as India, calling Mount Parapamisos the Kaukasos, for the sake of Alexander's reputation, so that Alexander would have crossed the Kaukasos. (4) In India itself, when they saw cattle branded with a club this was proof that Herakles had come to India. Eratosthenes is similarly disbelieving about the wanderings of Dionysos.

24 (IIC23). Strabo, Geography *11.7.4.*
Many false things were further imagined about this sea [the Hyrkanian] because of the ambition of Alexander. Since it was agreed by all that the Tanais River separated Asia from Europe, that which was between the sea and the Tanais, a greater part of Asia, had not fallen to the Makedonians. But it was reported in such a way as to show that Alexander had conquered that region. They made into one the Maiotic Lake (which receives the Tanais) and the Kaspian Sea, calling it a lake, and insisting that there was a passage from one to the other so that each was a part of the other. Polykleitos offers proofs that the sea is a lake (it produces serpents and the water is sweetish) and he judges that it is nothing other than the Maiotis because the Tanais empties into it. From the same Indian mountains that come the Ochos and Oxos and many others flows the Iaxartes, the most northerly of all, which like the rest empties into the Kaspian Sea. This they named the Tanais, and as an additional proof that it was the Tanais of which Polykleitos spoke, they note that across the river the fir tree exists and that the Skythians there use fir arrows. This is their proof that the territory across the river is part of Europe and not Asia, for upper and eastern Asia do not produce the fir tree. But Eratosthenes says that the fir tree also grows in India and that Alexander built his ships there out of fir. Eratosthenes attempts to reconcile many other such issues, but let what I have said about them be enough.

Book 2

Introduction

25 (IIA1). Strabo, Geography *1.4.1.*
In his second book, he [Eratosthenes] attempts to change the structure of geography and states his own assumptions, and if there is any further correction, there must be an attempt to provide it. To introduce mathematics and physics into the topic is well considered, and also the idea that if the earth is spherical, just as the cosmos, it is inhabited all around, as well as other such comments. But later writers do not agree as to whether it is as large as he has said, nor do they approve of his measurements. Yet in regard to the signs of phenomena for each of the inhabited regions, Hipparchos also makes use of his intervals for the meridian through Meroë, Alexandria, and Borysthenes, saying that they differ only slightly from the truth. In what follows about its shape, where he proves at length that the nature of the earth (along with its wet portions) is spherical, as are the heavens, he seems to be speaking irrelevantly, for brevity would be sufficient.

Methodology

26 (IIA2). Theon of Alexandria 394–5.
Eratosthenes showed, making use of dioptras—which measure from a distance—that a line falling from the higher mountains is ten stadia.

27 (IIB43). Pliny, Natural History *12.53.*
The *schoinos* by the calculation of Eratosthenes is 40 stadia, that is five miles, but others have made a single *schoinos* 32 stadia.

The Size and Shape of the Earth

28 (IIB39). Pliny, Natural History *2.247–8.*
(247) Eratosthenes (who was thorough in every area of learning, but especially skilled in these matters: I see that he is accepted by everyone)

recorded the complete circumference at 252,000 stadia, a calculation that by Roman measurement results in 31,500 miles, a presumptuous deed, but actually understood through such a subtle argument that one is ashamed not to believe. Hipparchos, who, in his refutation of him and in all his remaining carefulness is amazing, adds a little less than 26,000 stadia.

(248) Dionysodoros has another belief: I will not eliminate this great example of Greek foolishness. He was a Melian, and distinguished in the discipline of geometry. He met up with old age in his native land and his inheritance came to his female relatives who performed his funeral. It is said that as they were carrying out the ceremonies on the following days, they found in his tomb a letter with the name of Dionysodoros, written to those above, saying that he had gone from his tomb to the depths of the earth, and that it was 42,000 stadia. There was no lack of geometricians who interpreted the letter to signify that it had been sent from the center of the sphere of the earth, which was the longest distance down from the surface and was also the center of the sphere. From this the calculation followed that caused them to pronounce that the circumference was 252,000 stadia.

29 (IIB12). *Markianos of Herakleia,* Periplous of the External Sea *1.4* (GGM *vol. 1, p. 519).*
Eratosthenes of Kyrene says that the maximum circumference of the earth is 259,200 stadia. This is also the measurement of Dionysios son of Diogenes.

30 (IIB27). *Strabo,* Geography 2.5.5–6.
(5) Let us assume that the earth along with the sea is sphere shaped and that one and the same surface contains the ocean, for the projections on the earth would be concealed because they are small in comparison with its great size and would escape notice. Thus we call it "sphere shaped," not as if turned on a lathe nor as a surveyor would present, but in order to perceive it, and this somewhat roughly. Let us consider a five-zoned [earth], with the equator drawn as a circle on it, and another [circle] parallel to it bordering the cold region in the northern hemisphere, and another at right angles through the poles. Since the northern hemisphere contains two-fourths of the earth, made by the equator and that [line] passing through the poles, in each of them a four-sided area is cut off, of which the northern side is half of the parallel next to the pole, the southern side is half the equator, and the remaining sides are sections of those [lines] passing through the poles,

which lie opposite to each other and are equal in length. In either one of these four-sided areas (it would seem to make no difference which one) we say that our inhabited world is placed, washed all around by the sea and like an island. It has been said that this is shown through perception and reason. If anyone were not to believe this argument, it would make no difference to geography whether to make it an island or to admit what we understand from experience, that one can sail around both sides, from the east and the west, except for a few places in the middle. Regarding these, there is no difference whether they are bounded by sea or uninhabitable land, for the geographer attempts to speak about the known parts of the inhabitable world. He omits the unknown parts, as well as that which is outside of it. It will suffice to join with a straight line the farthest limits of the coastal voyage on both sides and to fill completely the form of the so-called island.

(6) Let us propose, then, that the island is in the previously mentioned quadrilateral. It is necessary to take as its size what it appears to be, removing our hemisphere from the entire size of the earth, and from this its half, and then also from it the quadrilateral in which we say that the inhabited world lies. It is necessary to understand its shape by analogy, adapting its appearance to the hypothesis. But since the segment of the northern hemisphere that is between the equator and the polar parallel is in the shape of a spindle whorl, and since the polar [parallel] cuts the hemisphere into two, and also cuts the spindle whorl in two, making the quadrilateral, it will be clear that the quadrilateral in which the Atlantic Ocean lies appears as half the spindle whorl, and that the inhabited world is a chlamys-shaped island in it, less in size than half the quadrilateral. This is clear from geometry, from the size of the sea that spreads around it (which covers the farthest point of the continent at both ends, contracting them to a tapering shape), and, third, from the extent of its length and width. The former is 70,000 stadia, limited for the most part by a sea that still cannot be crossed because of its size and desolation. The latter is less than 30,000 stadia, bounded by areas that are uninhabitable because of heat or cold. The part of the quadrilateral that is uninhabitable because of heat has a width of 8,800 stadia, and a maximum length of 126,000, which is half the equator . . . [there may be a lacuna in the text] and the remainder may be more.

31 (IIA7). Strabo, Geography *2.5.13.*
To describe accurately the entire earth and the whole "spindle whorl" of the zone of which we were speaking is another discipline, as is whether

the spindle whorl is inhabited in its other fourth portion. If it were, it would not be inhabited by those like the ones among us, and it must then be considered another inhabited world, which is believable. For myself, however, I must speak of what is in our own.

32 (IIC1). *Agathemeros 1.2.*

First Demokritos, a man of much experience, determined that the earth was elongated, with a length 1½ times the width, and Dikaiarchos the Peripatetic agreed. Eudoxos made the length twice the width, and Eratosthenes more than twice. . . .

33 (IIA6). *Strabo,* Geography *1.4.6–8.*

(6) He [Eratosthenes] attempts to reassure us even further when he says that it is natural to make the interval from east to west greater, and natural that the inhabited world is longer from east to west, saying that "we have stated, as do the mathematicians, that it joins together in a circle, touching itself with itself, so that, if not prevented by the size of the Atlantic Ocean, we could sail from Iberia to India along the same parallel (the part that remains beyond the previously mentioned interval, which is more than one-third of the distance), if the one through Athens is less than 200,000 stadia, where we have made this stated measurement of stadia from India to Iberia." Yet he does not say this well, for although he might say this about the temperate zone (ours), according to mathematics, since it is only a portion of the inhabited earth, yet concerning the inhabited earth—since we call inhabited that which we inhabit and know—it is possible that in this same temperate zone there are two inhabited worlds, or more, especially near the circle through Athens that is drawn through the Atlantic Ocean. Moreover, by persisting in a spheroid shape of the earth he happens upon the same criticism. Similarly he does not stop disagreeing with Homer about the same things.

(7) Next, saying that there has been much written about the continents, and that some divide them by rivers, such as the Nile and Tanais, representing them as islands, and others by isthmoi, such as the one between the Kaspian and Pontic Seas or between the Erythraian and the Ekregma, saying that they are peninsulas, he says that he does not see how this examination can result in anything consequential except for those living contentiously in the fashion of Demokritos. If there are no exact boundaries—as with Kolyttos and Melite—such as stelai or enclosures, we can only say that "this is Kolyttos and this is Melite," but do not have the boundaries. It is for this reason that there are often disputes about districts, such as that between the Argives and

Lakedaimonians about Thyrea, or between the Athenians and Boiotians about Oropos. The Hellenes named the three continents differently because they did not pay attention to the inhabited world but to only their own area and what was directly opposite, Karia, where there are now Ionians and their neighbors. In time, advancing still farther, and learning more about territories, they have focused their division. Whether those who first separated the three (so that we begin with his [Eratosthenes'] last point, living contentiously but not in the fashion of Demokritos, but of him): so did these original men seek to divide Karia, lying opposite, from their own territory? Or did they conceive only of Hellas and Karia and the small amount that touched, but not in the same way about Europe, Asia, or Libya, but those afterward, traveling through enough to conceive of the outline of the inhabited world: were these the ones who divided it into three? How could they not have made the division of the inhabited world? Would someone speaking of three parts and calling each of the parts a continent not think of the whole from which he makes his division? But if he does not conceive of the inhabited world but makes his division of some part of it, what part of the inhabited world would anyone have said Asia was a part, or Europe, or a continent in general? These things have been said sloppily.

(8) Even more sloppily, he does not see what can be said about the practical result of the investigation of boundaries, to set forth Kolyttos and Melite and then to turn around to the opposite. If the wars about Thyrea and Oropos happened because of ignorance of the boundaries, then the separation of territories results in something practical. Or is he saying, in regard to districts, and, by Zeus, the various ethnic groups, that it is practical to divide them accurately, but for continents this is superfluous? But this is by no means less important, for there might be some great dispute among rulers, one holding Asia and the other Libya, as to which possessed Egypt, specifically that which is called the lower territory of Egypt. If anyone were to dismiss this because of its rarity, nevertheless it must be said that the continents are divided according to a major distinction that relates to the entire inhabited world. In regard to this, there must be no concern that certain areas remain undefined, if the division is made according to rivers, because the rivers do not extend as far as the Ocean and thus truly do not leave the continents as islands.

34 (IIIA39, IIB15, IIB23). Strabo, Geography 2.5.7–9.
(7) He [Hipparchos] says, based on the size of the earth, as Eratosthenes has stated, that it is necessary to consider the inhabited world sepa-

rately, for it will not make a great difference in regard to the celestial phenomena for each inhabited region whether his or later measurements are used. Since, according to Eratosthenes, the equator is 252,000 stadia, one fourth would be 63,000. This is the distance from the equator to the pole, fifteen sixtieths of the sixty [intervals] of the equator. From the equator to the summer tropic is four [sixtieths], and this is the parallel drawn through Syene. Each of these distances is computed from known measurements. The tropic lies at Syene because there at the summer solstice a gnomon has no shadow in the middle of the day. The meridian through Syene is drawn approximately along the course of the Nile from Meroë to Alexandria, which is about 10,000 stadia. It happens that Syene lies in the middle of that distance, so that from there to Meroë is 5,000. Going in a straight line about 3,000 stadia to the south, it is no longer inhabitable because of the heat, so the parallel through these regions, the same as the one through the Kinnamomophoroi region, must be put as the limit and the beginning of our inhabited world in the south. Since it is 5,000 from Syene to Meroë, adding the other 3,000 the total is 8,000 to the boundary of the inhabited world. But from Syene to the equator is 16,800 (this is the four sixtieths, with each 4,200), thus the remainder would be 8,800 from the boundary of the inhabited world to the equator, and 21,800 from Alexandria. Again, everyone agrees that the sea route to Rhodes is in line with the course of the Nile, as well as the sailing route from there along Karia and Ionia to the Troad, Byzantion, and the Borysthenes. Taking, then, the known distances that have been sailed, they consider how far the territories in a straight line beyond the Borysthenes are inhabitable and what is the boundary of the part of the inhabited world toward the north. The Roxolanoi, the farthest of the known Skythians, live beyond the Borysthenes, although they are farther south than the remote peoples we know about north of Brettanike. The area lying beyond immediately becomes uninhabitable because of the cold. Farther to the south of them are the Sauromatai beyond the Maiotis and the Skythians as far as the eastern Skythians.

(8) Now Pytheas of Massalia says that the region around Thoule, the most northerly of the Prettanikai, is farthest, and that the circle of the summer tropic is the same as the arctic [circle].

[Strabo's objections to this omitted.]

With the parallel through Byzantion going approximately through Massalia, as Hipparchos said, believing Pytheas (he says that at Byzantion the relationship of the gnomon to its shadow is the same as that Pytheas reported for Massalia), and the one running through Borysthenes about 3,800 from there, considering the distance from Massalia the circle through the Borysthenes would fall somewhere in Prettanike.

[Strabo's further dismissal of Pytheas omitted.]

(9) If one were to add to the distance from Rhodes as far as Borysthenes a distance of 4,000 stadia from Borysthenes to the northern regions, this is a total of 12,700 stadia, and that from Rhodes to the southern limit of the inhabited world is 16,600, so the entire width of the inhabited world is less than 30,000 from south to north. The length is said to be about 70,000, that is, from west to east, from the extremities of Iberia to the extremities of India, measured in part by land journeys and in part by sea journeys. That this length is within the quadrilateral mentioned is clear from the relationship of the parallels to the equator, and thus the length is more than twice the width. It is said to be somewhat chlamys shaped, for when we travel throughout its regions, a great contracting of width at the extremities, especially at the west, is found.

35 (IIC2). Strabo, Geography 1.4.2.
In determining the width of the inhabited world, he [Eratosthenes] says that from Meroë it is 10,000 stadia along its meridian to Alexandria, and from there to the Hellespont about 8,100, and then 5,000 to Borysthenes, and then to the parallel that runs through Thoule (which Pytheas says is six days' sail north of Brettanike and is near the frozen sea) an additional 11,500. Moreover, if we add 3,400 more beyond Meroë, so that we include the Egyptian island, the Kinnamomophoroi, and Taprobane, we have 38,000 stadia.

36 (IIC5). Strabo, Geography 2.5.42.
Eratosthenes says that these regions [around the mouth of the Borysthenes] are a little more than 23,000 stadia from Meroë, since it is 18,000 stadia to the Hellespont and then 5,000 to Borysthenes.

37 (IIC18). Strabo, Geography 1.4.5.
Since he [Eratosthenes] entirely missed its width, he was also compelled to miss its length. It is agreed by later sources as well as the most talented early ones that its known length is more than twice the known width (I am speaking of that from the extremities of India to the extremities of Iberia, and that from Aithiopia as far as the parallel of Ierne). He has determined the previously mentioned width, that from farthest Aithiopia as far as the parallel through Thoule, and has stretched the length more than necessary, so that he can make it more than the previously mentioned width. Moreover, he says that the narrowest part

of India, up to the Indos River, is 16,000 stadia—that which extends to the promontories extends an additional 3,000 stadia—and to the Kaspian Gates is 14,000, and then to the Euphrates 10,000, and from the Euphrates to the Nile 5,000, with an additional 1,300 as far as the Kanobic mouth, then 13,500 as far as Karchedon, then as far as the Pillars at least 8,000, a total of 800 beyond 70,000 stadia. Then it is still necessary to add the bulge of Europe outside the Pillars of Herakles, set against the Iberians and sloping to the west—no less than 3,000 stadia—as well as all the promontories, especially that of the Ostimioi, which is called Kabaion, and the surrounding islands, the farthest of which, Ouxisame, Pytheas says is three days' sail away. After mentioning these final places, which in their extent add nothing to its length, he mentions the districts around the promontories, that of the Ostimioi and Ouxisamai and their islands. All these places are toward the north and are in Keltika, not Iberika, or rather they are fantasies of Pytheas. He [Eratosthenes] also adds to the previously mentioned length more stadia, 2,000 to the west and 2,000 to the east, to keep the width from being more than half the length.

38 (IIC19). Measurement of the Entire Inhabited Earth 1
(GGM *vol. 1, p. 424*).
The length of our inhabited earth from the mouth of the Ganges to Gades is 83,800 stadia, and the width from the Aithiopian Sea to the Tanais River is 35,000 stadia. The area between the Euphrates and the Tigris Rivers, called Mesopotamia, is a distance of 3,000 stadia. Eratosthenes, one of the most learned men of antiquity, made this measurement.

The Nature of the Ocean

39 (IIA13). *Strabo*, Geography 1.1.8–9.
(8) That the inhabited world is an island must be assumed first from the senses as well as experience. Everywhere that it has been possible for men to access the farthest points of the earth, the sea has been found, which we call Okeanos. Wherever it is not possible to make use of the senses, reason shows it. The eastern side—that around India—and the western—that around Iberia and Maurousia—can be completely sailed around for a great distance in the southern or northern portions. The remainder that has not been sailed by us up to today, because those who sailed around did not meet each other, is not so great, if one adds together the parallel distances accessible by us. It is not likely that the Atlantic Ocean is divided into two seas, separated by isthmoi so narrow

that they prevent sailing around, but rather that it flows together and is continuous. Those who attempted to sail around but turned back say that it was not because they came upon some continent and were prevented from sailing beyond, reversing their direction, but because of difficulties and isolation, not because of a lessening of the sea, which was still passable. This agrees better with the properties of the Ocean, concerning its ebb and flooding. Everywhere the same characteristic—or one not greatly varying—is enough for the changes of height and diminution, as if one sea and one cause produced the movements.

(9) Hipparchos is not believable in refuting this idea on the grounds that the Ocean is not affected in the same way everywhere, and, even if this were so, it does not follow that the Atlantic Ocean flows around in a complete circle, citing Seleukos of Babylon as witness that it is not affected in the same way.

The Unique Qualities of Syene and Its Region

40 (IIB36). Strabo, Geography *2.1.20.*
Regarding the latitude of Meroë, Philon, who wrote about his voyage to Aithiopia, records that the sun is at the zenith 45 days before the summer solstice, and also discusses the relationship of the gnomon to the shadows of both the solstices and equinoxes. Eratosthenes is closely in agreement with Philon.

41 (IIB38). Pliny, Natural History *2.183–5.*
(183) Similarly, they say that in the town of Syene, which is 5,000 stadia above Alexandria, no shadow is cast at noon on the day of the solstice, and that a well made in order to test this is totally illuminated, which shows that the sun is vertically above that place, at that time. Onesikritos wrote that this also occurs at the same time in India above the Hypasis River. It is also established that at Berenike, the Trogodyte city, and 4,820 stadia from there in the town of Ptolemais, belonging to the same people, which was founded on the shore of the Red Sea for the first elephant hunts, the same thing happens 45 days before and after the solstice, and during those 90 days the shadows are thrown to the south. (184) Moreover, in Meroë (an island inhabited in the Nile River and the capital of the Aithiopian people, 5,000 stadia from Syene), shadows are absent twice a year, when the sun is in the eighteenth part of Taurus and the fourteenth of Leo.

[Solar phenomena observed by those with Alexander omitted.]

(185) Eratosthenes states that in all of Trogodytika the shadows fall the wrong way twice a year for 45 days.

42 (IIB37). Pliny, Natural History *6.171.*
This is the region indicated by us in Book 2 [183], where for 45 days before the solstice and for the same period afterward, the shadows are reduced to nothing at the sixth hour, and for the rest of the day fall to the south, but toward the north on the remaining days, while at Berenike— the first one we mentioned—on the day of the solstice itself the shadow is completely absent at the sixth hour, but nothing else unusual is observed. This is 602½ miles distant from Ptolemais. This is matter of great significance and infinite perception, where the world was grasped, because there Eratosthenes conceived of the measurement of the earth by the unquestioned behavior of the shadows.

43 (IIB40). Ammianus Marcellinus 22.15.31.
Then there is Syene, where, at the time of the solstice, the sun extends its summer path and its rays surround everything upright and do not allow shadows to go beyond the bodies themselves. At that time, if one fixes a stick upright, or looks at a standing man or a tree, he will observe that shadows are absent around the extremities of the outlines of the figures. Thus it is at Meroë, a part of Aithiopia near the equinoctial circle, where, it is said, for 90 days the shadows fall opposite to ours, and because of which the inhabitants are called Antiskioi.

The Terrestrial Zones

44 (IIB26). Geminos, Introduction to the Phenomena *15.*
The surface of the earth is sphere shaped and divided into five zones. Of these, the two around the poles, lying farthest from the path of the sun, are called the chilled ones and are uninhabitable because of the cold, bounded by the arctic [circles and extending] toward the poles. The next ones, which lie in a moderate position relative to the path of the sun, are called temperate. They are bounded by the celestial arctic and tropical circles and lie between them. The remaining one, in the middle of those previously mentioned and which lies under the path of the sun, is called burned. It is cut in two by the circle of the equator of the earth, which lies under the circle of the celestial equator. In regard to the two temperate zones, the boreal corresponds to the inhabited one in which we live, whose length is about 100,000 stadia, and whose width is about half.

45 (IIA5). Strabo, Geography *2.3.2.*

If, as Eratosthenes says, that which lies under the celestial equator is temperate—and Polybios agrees with this opinion, although he adds that it is the highest part, and because of this it is rainy, since in the Etesian season the clouds from the north frequently strike against the heights there—it would be much better to consider it a third, narrow, temperate zone, than to introduce two tropical ones.

Book 3

The Plan of the Inhabited World

46 (IIIA24). Strabo, Geography *2.5.16.*
Such being the shape of the entire [inhabited world], it appears useful to take two straight lines, which cut across each other at a right angle, one going through all the greatest width and the other the length, and the first will be one of the parallels and the other one of the meridians. Then one should think of lines parallel to these on either side, which are used to divide the land and the sea that we happen to use. Thus the shape will be somewhat more clear, as I have described, according to the length of the line, with different measurements for both the length and width, and the terrestrial regions will be better manifested, both in the east and west as well in as the south and north.

47 (IIIA2). Strabo, Geography *2.1.1–3.*
(1) In the third book of the *Geographika*, establishing the plan of the inhabited world, he [Eratosthenes] divides it into two parts by a certain line from west to east, parallel to the line of the equator. He takes as the extremities of this line in the west the Pillars of Herakles and in the east the farthest summits of the mountains that define the northern edge of India. He draws the line from the Pillars through the Sikelian strait and the southern summits of the Peloponnesos and Attika, and as far as Rhodes and the Issic Gulf. Up to here, he says, the previously mentioned line runs through the sea and the adjacent land (in fact, it entirely lies along the length of our sea as far as Kilikia). Then it is thrown out as an approximately straight line along the entire Tauros mountain range as far as India, for the Tauros runs in a straight line with the sea from the Pillars, dividing Asia lengthwise into two parts, making one the northern part and the other the southern: thus in a similar way it [the Tauros] lies on the parallel through Athens, as does the sea that comes as far as it from the Pillars.

(2) Having said this, he believes it necessary to correct the ancient geographical plan, for according to it the eastern portion of the mountains is greatly twisted toward the north, and India is drawn along further to the north than it should be. He offers as his primary proof that the most southerly promontories of India rise opposite to the region around Meroë, as most agree, demonstrated by the climate and the celestial phenomena. From there [southern India] to the northern part of India, that at the Kaukasos Mountains, Patrokles—who is believed to be most accurate both because of his reputation and because he is not uneducated as a geographer—says is 15,000 stadia, which is about the same as from Meroë to the parallel through Athens, and thus the northern parts of India, which touch the Kaukasos mountains, end at this latitude.

(3) Another proof he offers is that the distance from the Issic Gulf to the Pontic Sea is about 3,000 stadia, going toward the north and the regions around Amisos and Sinope, equal to what is said for the width of the mountains. From Amisos, heading toward the equinoctial sunrise, there is first Kolchis and then the pass to the Hyrkanian Sea, and next the route to Baktra and the Skythai beyond (having the mountains on the right). This line, through Amisos [extended] to the west, is thrown out through the Propontis and the Hellespont. From Meroë to the Hellespont is no more than 18,000 stadia, as much as from the southern side of India to the parts around the Baktrians, adding 3,000 to the 15,000, part of which is due to the width of the mountains and part due to India itself.

48 (IIIA23). Strabo, Geography *11.12.4–5.*
(4) Thus I place Media, in which the Kaspian Gates are, within the Tauros, as well as Armenia. (5) According to me, then, these peoples would be toward the north, but Eratosthenes, who made the division into a southern part and a northern, calling some of his previously mentioned "sealstones" northern and others southern, declares the Kaspian Gates to be the boundary between the two latitudinal regions. Reasonably, he would declare the southern part that which is more southerly, stretching toward the east, rather than the Kaspian Gates, among which are Media and Armenia, and the northern part the more northerly, since this happens regardless of the distribution of sections. But perhaps it did not occur to him that no part of either Armenia or Media is south or outside of the Tauros.

49 (IIIB3, IIIB7). Strabo, Geography *2.1.31.*
He [Eratosthenes] has cheerfully divided the inhabited world into two parts by means of the Tauros and the sea to the Pillars. In regard to

the southern portion, the Indian borders have been described well in terms of a mountain, river, and sea, and by a single name, that of a single people, so that he [Eratosthenes] correctly calls it four-sided and rhomboidal.

50 (IIIA8). Strabo, Geography *2.1.5.*
First, although he [Eratosthenes] used many testimonia, he [Hipparchos] says that only one was used, Patrokles. But who said that the southern promontories of India rise opposite to the regions of Meroë? Who said that from Meroë as far as the parallel of Athens was such a distance? Again, who said what the width of the mountains was, and that from Kilikia to Amisos was the same? Who said that from Amisos through Kolchis and Hyrkania as far as Baktria and the regions beyond down to the eastern sea was in a straight line toward equinoctial east, along the mountains that are on the right? Or again that toward the west was straight with this line toward the Propontis and Hellespont? Eratosthenes takes all these things as established by those who had been in those places, for he studied many treatises, having them in abundance in a library as large as Hipparchos says it was.

51 (IIIA11, IIIA35). Strabo, Geography *2.1.10–11.*
(10). . . if the meridian through Rhodes and Byzantion has been taken correctly, then the one through Kilikia and Amisos has also been taken correctly, since from many sources it is shown that lines are parallel if neither of them meets.

(11) Sailing from Amisos to Kolchis is thus toward equatorial east, which is shown by the winds, seasons, crops, and the sunrise itself, as also is the pass to the Kaspian and the route from there to Baktra. Often clarity and total agreement are more trustworthy than an instrument. Hipparchos himself, in regard to the line from the Pillars as far as Kilikia—that it was straight and toward equinoctial east—did not totally make use of instruments and geometry, but for its entirety from the Porthmos to the Straits he trusted sailors, so he is not accurate in saying, "We cannot say in regard to either the relation of the longest day to the shortest, or the gnomon to its shadow along the mountainside from Kilikia as far as India, whether the slant is a parallel line, but it must not be corrected, preserving the slant that the ancient maps have." First, "cannot say" is the same as withholding it, and the one withholding inclines neither way. When he exhorts that it be left alone, just like the ancients, he inclines to neither. Rather he would be preserving his

consistency if he advised not to use geography at all, for we "cannot say" what the positions are of the other mountains, such as the Alpes, Pyrenaioi, the Thracian, Illyrikan, or Germanikan. Who would believe that the ancients were more trustworthy than those more recent, since they made all those mistakes in drawing plans that Eratosthenes has accused them of, none of which Hipparchos objected to.

52 (IIIA14, IIIA34, IIIB65). Strabo, Geography 2.1.39.
He [Hipparchos] is also mistaken in his next undertaking, in which he wishes to conclude that the route from Thapsakos to the Kaspian Gates (which Eratosthenes said is 10,000 stadia recorded as a straight line) was not measured in a straight line, although the straight line is much shorter. His attack on him is as follows: he says that according to Eratosthenes himself the meridians through the Kanobic mouth and through the Kyaneai are the same, and this is 6,300 stadia from the one through Thapsakos, while the Kyaneai are 6,600 from Mt. Kaspios, which lies at the pass to the Kaspian Sea from Kolchis. Thus the distance from the meridian through the Kyaneai to Thapsakos is within 300 stadia of the same as to Kaspios, and thus, essentially, Thapsakos and Kaspios are on the same meridian. Therefore it follows that the Kaspian Gates are equidistant from Thapsakos and Kaspios, and Kaspios is much closer to the gates than the 10,000 that Eratosthenes says they are from Thapsakos. Therefore the distance from Thapsakos is much less than the 10,000 of the straight line, and thus the 10,000 that he measures on a straight line from the Kaspian Gates to Thapsakos is circuitous. I say to him [Hipparchos] that Eratosthenes makes his line loosely, as is proper in geography, and also makes his meridians and lines to the equinoctial east loosely, but he [Hipparchos] critiques them geometrically, as if each had been drawn with instruments. Yet he does not use instruments himself, rather taking [the relationship of the perpendicular and parallel] by guessing. This is one of his mistakes. Another is that he does not put down the measurements that were produced [by Eratosthenes] or put them to the test, but only those created by himself. Thus, although he [Eratosthenes] first said that this distance from the mouth [of the Pontos] to Phasis was 8,000 stadia, and added the 600 on to Dioskourias, and then five days' crossing over to Kaspios (which Hipparchos represents as 1,000 stadia), so that in all it totals, according to Eratosthenes, 9,600 stadia, he makes a shortcut and says that from the Kyaneai to Phasis is 5,600 and from there to Kaspios another 1,000. Thus it is not according to Eratosthenes that Kaspios

and Thapsakos are essentially on the same meridian, but according to him [Hipparchos].

The Shape of the Inhabited World

53 (IIIA12). Strabo, Geography *2.5.14.*
The shape of the inhabited world is somewhat in the form of a chlamys, whose greatest width is marked by the line through the Nile, with its beginning taken at the parallel through the Kinnamomophoroi and the island of the fugitive Egyptians as far as the parallel through Ierne. The length is at right angles, from the west through the Pillars and the Sikelian Strait as far as Rhodes and the Issic Gulf, going through the Tauros that girdles Asia and ends at the eastern sea between India and the Skythians beyond Baktriana. It is necessary to conceive of a certain parallelogram in which the chlamys-shaped form is engraved so that its greatest length agrees with and is equal to the length [of the parallelogram] and the width agrees with [and is equal to] its width. This chlamys-shaped form is the inhabited world. Its width, as we have said, is bounded by the farthest sides of the parallelogram, which separate the inhabited and uninhabited parts from one another. These are, on the north, that through Ierne, and in the burned region, that through the Kinnamomophoroi. Extending these to the east and west as far as the portions of the inhabited world that rise opposite to them, they make a certain parallelogram, joining up with those at the extremities. It is clear that the inhabited world is within this because neither the greatest width nor length falls outside, and that the form is chlamys shaped because the extremities of its length taper on both sides and diminish its width, washed away by the sea. This is clear from those who have sailed around the eastern and western portions on either side. They proclaim that the island of Taprobane is further south than India by far, but nonetheless inhabited, rising opposite to the Island of the Egyptians and the land of the Kinnamomophoroi, and that the temperature of the air is about the same. The regions around the mouth of the Hyrkanian Sea are farther north than the ultimate part of Skythia beyond India, and that around Ierne is still farther. Similar things are said about the region beyond the Pillars, that the most western boundary of the inhabited world is the promontory of Iberia called Hieron ["Sacred"], which lies approximately on the line through Gadeira, the Pillars, the Sikelian Strait, and Rhodes. The *oroskopeia* agree, they say, and the winds are favorable in both directions, and the lengths of

the longest days and nights [agree], for the longest days and nights have 14½ equinoctial hours.

54 (IIIA15). Strabo, Geography *2.1.35.*
He [Eratosthenes] declares that differences of 400 stadia can be perceived, such as that between the parallels of Athens and of Rhodes. Observing this is not something done by a single method, but there is one [method used] where the difference is greater and another [where] less. Where it is greater, we can trust our eyes or the crops, or the temperature of the air in judging the latitude, but for the lesser there are instruments such as sundials or dioptras. Thus when taking the parallels of Athens with a sundial, and that of Rhodes and Karia, the difference is perceptible, as is expected with so many stadia. But, in a width of 3,000 stadia and a length of 40,000 stadia in the mountains and 30,000 in the sea, when someone makes a line from west to equinoctial east, naming one part the northern and the other the southern, calling them "the rectangle" and "the sealstone," we must understand what he means by these terms, as well as "northern side" or "southern," and, moreover, "western" and "eastern." If he disregards this, he is greatly in error and must be held to account (for it is just), but if it is merely slight, even if he disregards it, he should not be questioned. In this there is no refutation to be made against him [Eratosthenes].

55 (IIIA16). Strabo, Geography *2.1.37.*
This is not where one must criticize Eratosthenes, but we do say that his loose magnitudes and figures must have some measurement, and that in some cases more must be conceded, in others less. Taking the width of the mountains that stretch toward the equinoctial east as 3,000 stadia, and similarly the sea as far as the Pillars [some elaboration by Strabo deleted] . . . or the width of the entire Tauros or the sea up to the Pillars as 3,000 stadia, we perceive a parallelogram that marks the outline of the entire mountain range and the previously mentioned sea.

[Further elaboration by Strabo deleted.]
But when he takes [the line] from the Kaspian Gates through the mountains themselves and the one that immediately diverges greatly from the mountains into Thapsakos, as if they led as far as the Pillars on the same parallel, and again throws out [a line] from Thapsakos as far as Egypt, taking in this width, and then measures the length of the figure by this length, he would seem to be measuring the lengths of the rectangle by the diagonal of the rectangle. When it is not a diagonal but a

deflected line, he would seem to err much more, for it is a deflected line leading from the Kaspian Gates through Thapsakos to the Nile.

56 (IIIB46). Strabo, Geography 2.1.33.
Eratosthenes takes the length of the inhabited world on the line through the Pillars, Kaspian Gates, and Kaukasos, as if straight, and that of the third section on that through the Kaspian Gates and Thapsakos, and that of the fourth section on that through Thapsakos and Heroonpolis as far as the region between the mouths of the Nile, which must come to an end in the region around Kanobos and Alexandria, for the last mouth is there, called the Kanobic or Herakleotic. Whether he places these lengths straight with each other or as if making an angle at Thapsakos, it is clear from what he says that neither is parallel to the length of the inhabited world. He draws the length of the inhabited world straight from the Tauros through the sea as far as the Pillars on a line through the Kaukasos, Rhodes, and Athens, and he says that from Rhodes to Alexandria along the meridian through them is not much less than 4,000 stadia: thus the parallels through Rhodes and Alexandria would be this [distance] apart from one another. That at Heroonpolis is about the same [as Alexandria], or somewhat farther south, and thus the line intersecting that parallel and that of Rhodes and the Kaspian Gates, whether straight or deflected, cannot be parallel to either. Thus he has not taken the lengths well, nor has he taken well the portions stretching northward.

The Parallels

57 (IIIA17). Strabo, Geography 2.5.35.
He [Hipparchos] says that those who live on the parallel through the Kinnamomophoroi—which lies 3,000 stadia from Meroë toward the south, and the equator is 8,800 from it—live very near to the midpoint between the equator and the summer tropic that passes through Syene, for Syene is 5,000 from Meroë. They [the Kinnamomophoroi] are the first for whom the Little Bear is completely within the arctic [circle] and is always visible, for the bright star at the extremity of the tail, the most southerly, is seated in the arctic circle so that it touches the horizon. The Arabian Gulf lies to the east and approximately parallel to the meridian discussed, and the Kinnamomophoroi are where it empties into the outer sea, where in ancient times they hunted the elephant. Its parallel extends on one side slightly to the south of Taprobane, or its farthest inhabitants, and on the other through southernmost Libya.

58 (IIB22). Strabo, Geography *2.2.2.*
From Syene, which is on the boundary of the summer tropic, to Meroë is 5,000 [stadia], and from there to the parallel of the Kinnamomophoroi, which is the beginning of the burned [zone], is 3,000. All of this distance is measurable by sea and by land, but the remainder as far as the equator, is shown to be 8,800 stadia, by means of the measurement of the earth made by Eratosthenes. The relationship of the 16,800 to the 8,800 would be [the relationship] of the distance between the tropics to the width of the burned [zone].

59 (IIIA18, IIIA19). Strabo, Geography *2.5.36.*
In the region of Meroë and Ptolemais—that among the Trogodytes—the longest day is 13 equinoctial hours, and this inhabited region is about midway between the equator and [the parallel] through Alexandria (it is 1,800 more to the equator). The parallel through Meroë passes on one side through regions that are unknown and on the other through the promontories of India. At Syene, at Berenike on the Arabian Gulf, and in Trogodytika the sun is in the zenith at the summer solstice, and the longest day is 13½ equinoctial hours, and almost the entire Great Bear is visible in the arctic [circle], except for the legs, the tip of the tail, and one of the stars in the square. The parallel through Syene passes, on one side, through the territory of the Ichthyophagoi in Gedrosia and through India, and on the other side through territory almost 5,000 stadia south of Kyrene.

60 (IIIA20, IIIA21, IIIA22). Strabo, Geography *2.5.38–41.*
(38) In the region about 400 stadia farther to the south of [the parallel through] Alexandria and Kyrene, where the longest day is 14 equinoctial hours, Arcturus is in the zenith, inclined a little toward the south. At Alexandria the relationship of the gnomon to the equinoctial shadow is five to three. This is 1,300 stadia farther south that Karchedon, if at Karchedon the relationship of the gnomon to the equinoctial shadow is 11 to seven. The parallel passes on one side through Kyrene and the region 900 stadia south of Karchedon and as far as the middle of Maurousia, and on the other side through Egypt, Koile Syria, Upper Syria, Babylonia, Sousiana, Persis, Karmania, and Upper Gedrosia, as far as India.

(39) In the region of Ptolemais—the one in Phoenicia—and Sidon and Tyre, the longest day has 14¼ equinoctial hours. These regions are about 1,600 stadia farther north than Alexandria and about 700 from Karchedon. In the Peloponnesos and around the middle of Rhodes,

around Xanthos in Lykia or a little to the south, and also 400 stadia south of Syrakousai, the longest day has 14½ equinoctial hours. These places are 3,640 from Alexandria, and, according to Eratosthenes, the parallel runs through Karia, Lykaonia, Media, the Kaspian Gates, and that part of India along the Kaukasos.

(40) In the area around Alexandria Troas, around Amphipolis, Apollonia in Epeiros, and south of Rome but north of Neapolis, the longest day has 15 equinoctial hours. The parallel is about 7,000 stadia north of the one through Alexandria in Egypt and more than 28,800 from the equator, 3,400 from the one through Rhodes, and 1,500 south of Byzantion, Nikaia, and the region around Massalia. Somewhat to the north is the one through Lysimacheia, which Eratosthenes says passes through Mysia, Paphlagonia, the region around Sinope, Hyrkania, and Baktra.

(41) In the region around Byzantion the longest day has 15¼ equinoctial hours and the relationship of the gnomon to its shadow and the summer solstice is 120 to 42 less a fifth. These places are 4,200 from [the parallel] through the center of Rhodes and about 30,300 from the equator. Sailing into the Pontos and going about 1,400 stadia to the north the longest day is 15½ equinoctial hours. These places are equidistant from the pole and the equatorial circle, and the arctic circle is in the zenith.

61 (IIIA13). Strabo, Geography *2.1.16.*
Eratosthenes cites this epigram in the temple of Asklepios at Pantikapaion, on a bronze hydria that had been broken by the frost:

> If any man does not believe what can happen here, let him look at this hydria and know, which has been presented by the priest Stratios, not as an honorable dedication to the god but as proof of our great winter.

Since [the climate] in the region around the Bosporos cannot be compared with that in the places enumerated, not even with Amisos and Sinope (which, we would say, are more temperate), it could hardly lie parallel to the region around Borysthenes and the farthest Kelts. It could scarcely be at the same parallel to the region around Amisos, Sinope, Byzantion, or Massalia, which are agreed to be 3,700 stadia farther south.

62 (IIIA28, IIA32, IIIA33, IIIB47). Strabo, Geography *2.1.36.*
He [Hipparchos] rightly objects, because he [Eratosthenes] called the line from Thapsakos to Egypt the length of this portion, as if one were

to say that the diagonal of a parallelogram is its length. For Thapsakos and the Egyptian coast do not lie on the same parallel, but ones far apart from one another, and the line from Thapsakos to Egypt is placed somewhat diagonally and at a slant between them. But when he [Hipparchos] is astonished at how confidently he [Eratosthenes] said that it is 6,000 stadia from Pelousion to Thapsakos, since it is more than 8,000 [7,000?], he is not correct.

[Hipparchos' arguments omitted.]

He [Eratosthenes] said that the route to Babylon from Thapsakos is 4,800 stadia and follows the Euphrates, this on purpose so that no one would take it as straight or a measurement of the distance between two parallels.

[Hipparchos' arguments omitted.]

Clearly he [Eratosthenes] has not granted that from Babylon to the Kaspian Gates is a meridian with a distance of 4,800.

63 (IIIA31, IIIB30, IIB17). Strabo, Geography 2.1.29.

And he [Hipparchos] says that if one considers a straight line from Thapsakos to the south, and one perpendicular to it from Babylon, there will be a right-angled triangle consisting of the side from Thapsakos to Babylon, the perpendicular leading from Babylon to the meridian line through Thapsakos, and the meridian itself through Thapsakos. He makes the hypotenuse of this triangle the line from Thapsakos to Babylon, which he says is 4,800 [stadia], and the perpendicular from Babylon to the meridian line through Thapsakos is slightly more than 1,000 stadia, as much as is the excess of [the line] to Thapsakos beyond that up to Babylon. From these he calculates the remaining [line] of the right angle to be much longer than the perpendicular mentioned. He adds to this that from Thapsakos to the north, running as far as the Armenian mountains, which Eratosthenes says had been measured at 1,100 stadia, but the rest is unmeasured and omitted. He [Hipparchos] assumes at least a thousand, so that both together are 2,100, and adding this to the straight side of the triangle, as far as the perpendicular from Babylon, he calculates a distance of many thousands from the Armenian mountains, or the parallel through Athens, as far as the perpendicular from Babylon, which he places on the parallel through Babylon—which he points out is no more than 2,400 stadia—assuming that the entire meridian is the number of stadia that Eratosthenes says. If so, the mountains of Armenia and those of the Tauros would not be on the parallel through Athens, as Eratosthenes [says], but many thousand stadia toward the north according to him. In addition, making further use of

the demolished assumptions about the structure of the right-angled triangle, he [Hipparchos] takes something not given, that the hypotenuse
in the right angle (the line straight from Thapsakos as far as Babylon)
is within 4,800 stadia. But Eratosthenes says that the route is along the
Euphrates, and also, saying that Mesopotamia along with Babylonia is
enclosed by the great circle of the Euphrates and Tigris, he says that
most of the circumference is due to the Euphrates. Therefore the straight
line from Thapsakos to Babylon would not be along the Euphrates or be
anywhere near as many stadia. Thus his [Hipparchos'] argument is destroyed. Moreover, I have already said that granting two lines drawn
from the Kaspian Gates, one to Thapsakos and the other to the part of
the Armenian mountains corresponding to Thapsakos, which Hipparchos himself has at least 2,000 stadia from Thapsakos, they could not be
parallel to each other or to that through Babylon, which Eratosthenes
calls the southern side. He said that the route along the mountain had
not been measured, but also said that from Thapsakos to the Kaspian
Gates [was measured], yet he added that one is speaking roughly. In addition, wishing only to speak about the territory between Ariana and
the Euphrates, there is no difference whether one or the other was
measured.

The Meridians

64 (IIIA27, IIIB11). Strabo, Geography *2.1.34.*
Then it follows that according to these assumptions [made by Hippar-
chos] the meridian line through the Kaspian Gates will intersect the
parallel through Babylonia and Sousa farther west than the intersection of the same parallel with the straight line running from the Kaspian Gates to the boundaries of Karmania and Persis, by more than
4,400 [stadia], and that this meridian line through the Kaspian Gates
would make almost half a right angle with the meridian from the Kaspian Gates to the boundaries of Karmania and Persis, and it will incline
midway between the south and equinoctial east. The Indos River will be
parallel to this, and thus it does not flow south from the mountains as
Eratosthenes says, but between that direction and equinoctial east, just
as it is marked on the old plans.
    [Comments about Hipparchos' errors deleted.]
He [Eratosthenes] has said that the shape of India is rhomboidal, and
just as its eastern side has been pulled far to the east, especially at the
farthest promontory, which is thrown to the south compared to the rest
of the coast, it is the same with the side along the Indos.

65 (IIIA40). Strabo, Geography *2.1.40.*
He [Eratosthenes] says it is 900 stadia [from Epidamnos to the Thermaic Gulf], and that from Alexandria to Karchedon is more than 13,000, although it is no more than 9,000 if Karia and Rhodes are on the same meridian as Alexandria and Porthmos on the same as Karchedon. All agree that the sail from Karia to Porthmos is no more than 9,000 stadia. In the case of a great interval, the meridian could be taken as the same as the more westerly one, that is, as far west as Karchedon is west of Porthmos, but in 4,000 stadia the error is manifest. And he places Rome on the same meridian as Karchedon, although it is more to the west, but he does not admit his excessive ignorance of these regions and of those on toward the west as far as the Pillars.

The Sealstone Concept

66 (IIIB2, IIIB5). Strabo, Geography *2.1.22.*
Following the thesis about the Tauros and the sea from the Pillars, he [Eratosthenes] divides the inhabited world into two parts by means of this line, calling them the northern part and the southern, and he attempts to divide each again into portions, insofar as possible, calling them "sealstones." He [Eratosthenes] says that the first sealstone of the southern portion is India, and the second Ariana, which are easy to sketch out, as he could render the width and length of both, but in a way to show the shape, as a geometrician. He says that India is rhomboidal because its sides are washed by the sea on the south and east, making shores without major gulfs, and the remainder by the mountains and the river, somewhat preserving there the rectilinear shape.

The First Sealstone (India)

67 (IIIA9). Strabo, Geography *2.1.19.*
Moreover, he [Eratosthenes] wishes to demonstrate that Deimachos is an amateur and inexperienced in such things, for he believes that India lies between the autumnal equinox and the winter tropic (contradicting Megasthenes, who says that in the southern portion of India the Bears are hidden and shadows fall in the opposite direction). But he [Deimachos] believes that neither occurs anywhere in India, and thus in asserting this he speaks with ignorance, for it is ignorant to think that the autumnal differs from the vernal in terms of its distance from the tropic, because the circle and the sunrise are the same at both. Since the distance between the tropic of the earth and the equator—where he

had placed India—has been shown through careful measurement to be much less than 20,000 stadia, this would turn out to be, even according to him [Deimachos], exactly what he [Eratosthenes] believes, not what the other believes, for if India were of that extent—or even 30,000—it could not fall within that distance, but if it is what he [Eratosthenes] says, it would fall within it. It is the same ignorance to say that nowhere in India do the Bears set or the shadows fall in the opposite direction, since it begins to happen as soon as one gets 5,000 from Alexandria. Again Hipparchos is not right in correcting what he [Eratosthenes] says.

68 (IIIA10). Strabo, Geography *2.1.20.*
No one records the latitudes of India, not even Eratosthenes. If, as they [Eratosthenes and Philon] think, both Bears set there, trusting those around Nearchos, then it is impossible that both Meroë and the Indian promontories are on the same parallel. If Eratosthenes agrees with those asserting that both the Bears set, why is it that no one reports on the latitudes in India, not even Eratosthenes? For this account is about latitude. If he does not agree with them, let him be discharged from the accusation. And he does not agree, for when Deimachos says that nowhere in India the Bears are hidden and the shadows fall in the opposite direction—which Megasthenes does in fact assume—he [Eratosthenes] charges him [Deimachos] with ignorance, and the entire combination is false.

69 (IIIB6). Strabo, Geography *15.1.10–11.*
(10) From my former discussion it especially seems most trustworthy what was expounded by Eratosthenes in the summary of the third book of his *Geographika* concerning what was then believed to be India, when Alexander invaded. The Indos was the boundary between it and Ariana, which is to the west and was then a Persian possession. Later the Indians held much of Ariana, having taken it from the Makedonians. This is what Eratosthenes says about this:
    (11) India is bounded on the north, from Ariana to the eastern sea, by the extremities of the Tauros, whose various parts the inhabitants call the Paropamisos, Emodos, and Imaos, and other names, but the Makedonians call it the Kaukasos. On the west is the Indos River, and the southern and eastern sides, much larger than the others, thrust into the Atlantic Ocean, and thus the shape of the territory becomes rhomboidal, with each of the larger sides having the advantage over the opposite sides by 3,000 stadia, as much as the promontory common to the

eastern and southern coast thrusts out equally on either side beyond
the remaining shore. The western side from the Kaukasos Mountains to
the southern sea is said to be about 13,000 stadia, along the Indos River
to its outlets. Thus the opposite—eastern—side, adding the 3,000 to the
promontory, will be 16,000 stadia. These are the least and greatest
widths of the territory. The length is from west to east, and one may
speak with certainty about it as far as Palibothra, for it has been measured with lines, and there is a royal road of 10,000 stadia. Beyond it is
known only by guess, by means of the voyage from the sea on the Ganges River as far as Palibothra, which would be about 6,000 stadia. The
entire extent, at least, is 16,000 stadia, which Eratosthenes says he took
from the most trusted record of stopping points. Megasthenes agrees,
although Patrokles says 1,000 less. The fact that the promontory thrusts
out farther east into the sea adds to this distance, and these 3,000 stadia make the greatest length. This is how it is from the outlets of the
Indos River along the successive shore to the previously mentioned
promontory and the eastern boundary. Those living there are called the
Koniakoi.

70 (IIIB8). Pliny, Natural History *6.56.*
. . . the Hemodi [Emodos] Mountains rise up, and the Indian peoples
begin, who not only live near the Eastern Ocean but also the Southern,
which we call the Indian. The part facing the east extends in a straight
line to a bend, and at the beginning of the Indian Ocean it totals 1,875
miles, and from there to where it bends to the south is 2,475 miles up to
the Indos River, which is the western border of India, as Eratosthenes
records.

71 (IIIB9). Arrian, Anabasis *5.6.2–3.*
(2) Dividing the southern part of Asia into four, Eratosthenes and Megasthenes (who lived with Sibyrtios the satrap of Arachosia and often
says that he visited Sandrokottos the Indian king) make the largest
portion the Indian land. The smallest is that bounded by the Euphrates
River, toward our interior sea. The other two are bounded by the Euphrates River and the Indos, which two together are hardly worthy of
comparison with the Indian land. (3) The territory of India toward the
east and the east wind is bounded to the south by the Great Ocean,
bounded on the north by Mt. Kaukasos as far as its connection with the
Tauros, and on the west and the Iapygian wind by the Indos River as far
as where the Great Sea cuts it off. Most of it is a plain, which, they believe, was deposited by rivers.

72 (IIIB10). Arrian, Indika *3.1–5.*
(1) For me, Eratosthenes of Kyrene is more trustworthy than anyone else, since the entire circuit of the earth was of interest to Eratosthenes. (2) He says that from Mt. Tauros, where the source of the Indos is, going along the Indos River itself to the Great Ocean and the outlets of the Indos, the side of the Indian land is 13,000 stadia. (3) The opposite side from the same mountain to the Eastern Ocean is more than equal to this side, since there is a promontory there that runs far out into the sea, and this promontory extends about 3,000 stadia. Thus he would make this side of the Indian land on the east 16,000 stadia. (4) This is what he determines as the width of the Indian land. Its length, from west to east as far as the city of Palimbothra, he says is recorded as measured in *schoinoi*, for there is a royal road, which extends for 10,000 stadia. (5) Beyond there it is not as precise. Those who record common reports say that along with the promontory that thrusts into the sea its extent is more than 10,000 stadia, but the width of the Indian land is more than 20,000 stadia.

73 (IIC21). Strabo, Geography *2.1.7.*
Moreover, Hipparchos records in his second book that Eratosthenes himself questions the reliability of Patrokles, because of his disagreement with Megasthenes about the length of India on its northern side, which Megasthenes says is 16,000 stadia, but Patrokles attests is a thousand less. Relying on a certain record of stopping points, he [Eratosthenes] distrusts them both because of their disagreement, holding to the record.

74 (IIIB12). Strabo, Geography *15.1.13–14.*
(13) All of India is watered by rivers, some of which flow into the two largest, the Indos and the Ganges, and others empty into the sea through their own mouths. All begin in the Kaukasos and run first toward the south, and although some continue in the same direction, especially those that join the Indos, others turn toward the east, such as the Ganges River. It flows down from the mountains and when it reaches the plain it turns toward the east and flows past Palibothra, a very large city, and then continues toward the sea and a single outlet, and is the largest of the Indian rivers. The Indos empties by two mouths into the southern sea, encompassing the land called Patalene, similar to the Egyptian Delta. It is because of the rising of vapors from these rivers and the Etesian winds, as Eratosthenes says, that India is inundated by summer rains and the plains become lakes. At the time of the rains, flax

and millet are sown, and in addition sesame, rice, and *bosmoron*. In the winter season there are wheat, barley, pulse, and other edible crops unused by us. Almost the same animals appear in Aithiopia and throughout Egypt as in India, and there are the same ones in the Indian rivers except the hippopotamus, although Onesikritos says that this horse is also there. The people in the south are the same as the Aithiopians in color, but in regard to eyes and hair they are like the others (because of the moisture in the air their hair is not curly). Those in the north are like the Egyptians.

(14) They say that Taprobane is an island in the Ocean, seven days' sail to the south from the southernmost part of India, the territory of the Koniakoi. Its length is 8,000 stadia, toward Aithiopia, and it has elephants. This is what Eratosthenes reports.

75 (IIIB17). Strabo, Geography *15.1.20.*
Megasthenes demonstrates the prosperity of India through its two yearly harvests and crops, as Eratosthenes also says, who mentions the winter sowing and that of the summer, as well as the rain. He says that no year is found to be without rain in both seasons, resulting in prosperity with the earth never barren. There are many fruit trees and plant roots, especially the large reeds that are sweet both naturally and when boiled, since the water is warmed by the sun, both that falling on account of Zeus and in the rivers. In a way, then, he wishes to say that what among others is called ripening—whether of fruits or juices—they call heating, and this is as effective as using fire to produce a good taste. In addition, he says that the branches of trees used in wheels are flexible, and for the same reason wool blooms on some.

76 (IIIB18). Pliny, Natural History *6.81.*
Eratosthenes puts forth the measurements [of Taprobane] as 7,000 stadia in width and 5,000 in length, and it contains no city but 700 villages.

The Second Sealstone (Ariana)

77 (IIIB22). Strabo, Geography *15.2.1.*
Ariana is after India, the first portion subject to the Persians after the Indos River and the upper satrapies outside the Tauros, bounded on the south and north by the same sea and the same mountains as India, and the same river, the Indos, from which it stretches toward the west as far as the line drawn from the Kaspian Gates to Karmania, so that its shape

is a quadrilateral. The southern side begins at the outlets of the Indos and at Patalene and ends at Karmania and the mouth of the Persian Gulf, having a promontory stretching considerably to the south, and then making a turn into the gulf toward Persis. First there are the Arbies (named like the Arbis River, the boundary between them and the next people, the Oreitai), who have a seacoast of 1,000 stadia, as Nearchos says. This is a part of India. Then there are the Oreitai, an autonomous people. The sail along the seacoast is 7,800 stadia, and along the next peoples, the Ichthyophagoi, 1,400, and along the Karmanian territory as far as Persis, 3,700, so that the total is 12,900.

78 (IIIB20, IIIB23). Strabo, Geography 15.2.8–9.
(8) It [Ariana] is large, and Gedrosia extends into the interior until it touches the Drangai, Arachotoi, and Paropamisadai, concerning which Eratosthenes has spoken as follows (for I cannot say it any better). He says that Ariana is bordered on the east by the Indos, on the south by the Great Ocean, on the north by the Paropamisos and the mountains continuing up to the Kaspian Gates, and on the west by the same boundaries that separate Parthyene from Media and Karmania from Paraitakene and Persis. The width of the territory is the length of the Indos from the Paropamisos to its outlets, 12,000 stadia (although some say 13,000), and the length from the Kaspian Gates is recorded in the treatise *Asiatic Stopping Points* in two ways. As far as Alexandria of the Areioi, from the Kaspian Gates through the Parthyaiai there is one route, and then there is a straight route through Baktriana and over the mountain pass into Ortospana to the meeting of three roads from Baktra, which is among the Paropamisadai. The other route turns slightly from Aria toward the south to Prophthasia in Drangiane, and the rest of it then goes back to the Indian boundary. This route through the territory of the Drangai and Arachosia is longer, 15,300 stadia in its entirety. If one were to remove 1,300, the remainder would be a straight line and the length of the territory would be 14,000. The seacoast is not much less, although some increase it and in addition to the 10,000 add Karmania with 6,000, including the gulfs or the seacoast of Karmania within the Persian Gulf. The name Ariana is extended to a certain part of Persis and Media as well as to the Baktrians and Sogdaians toward the north, who speak roughly the same language, only slightly different.

(9) The arrangement of the peoples is as follows: along the Indos are the Paropamisadai, above whom is Mt. Paropamisos. Then, toward the south, are the Arachotoi, and next toward the south the Gedrosenoi

along with the others on the seacoast, with the Indos lying alongside all these. Some of these places along the Indos are possessed by certain Indians, but were formerly Persian. Alexander took them away from the Arians and established his own foundation, and Seleukos Nikator gave them to Sandrakottos, concluding an intermarriage and receiving 500 elephants in return. Lying to the west of the Paropamisadai are the Arioi, and the Drangai [are west] of the Arachotoi and Gedrosioi, but the Arioi also lie to the north of the Drangai, as well as to the west, almost encircling a small part of them. Baktiane lies to the north of Aria and then there are the Paropamisadai, through whom Alexander passed over the Kaukasos pushing toward Baktra. To the west, next to the Arioi, are the Parthyaioi and the territory around the Kaspian Gates, and to the south is the Karmanian desert, and then the rest of Karmania and Gedrosia.

79 (IIIB19). Strabo, Geography *2.1.22.*
He [Eratosthenes] sees that Ariana has three sides suitably formed for the creation of a parallelogram, although he cannot mark off the western side by points, because the peoples there alternate with one another, yet he indicates it nevertheless by a line from the Kaspian Gates ending at the promontories of Karmania that touch the Persian Gulf. He calls this side the western and that along the Indos the eastern, but he does not say that they are parallel, nor the others, the ones delineated by the mountain and by the sea, but merely [calls them] the northern and the southern.

80 (IIIA30). Strabo, Geography *2.1.28.*
Eratosthenes has not said that the line bounding the western side of Ariana lies on a meridian, nor that [the line] from the Kaspian Gates to Thapsakos is at right angles with the meridian through the Kaspian Gates, but rather [mentions the line] marked by the mountain, with which [the line] from Thapsakos makes an angle, since it has been brought down from the same point as that from which the line at the mountain [has been drawn]. Moreover, he has not said that the line to Babylon from Karmania is parallel to the line to Thapsakos.

81 (IIIB24). Strabo, Geography *15.2.14.*
Karmania is the last place on the seacoast [that runs] from the Indos, although much farther north than the outlet of the Indos. Its first promontory, however, stretches to the south, toward the Great Ocean, mak-

ing the mouth of the Persian Gulf, along with the promontory extending from Arabia Eudaimon (which is in view), and it bends toward the Persian Gulf until it touches Persis.

The Third Sealstone (Mesopotamia)

82. Strabo, Geography *2.1.31.*
Ariana cannot easily be outlined because its western side is confused, but it is bounded by three sides, which are approximately straight, and also by its name, that of one people. But, as it has been determined, the third sealstone is completely undefined, for the common side between it and Ariana is confused, as I have said, and the southern side has been taken most sloppily, for it does not outline the sealstone, since it runs through the middle of it, and many portions toward the south are left out, nor does it trace the greatest length, for the northern side is longer. The Euphrates is not the western side, even if it flowed in a straight line, since its extremities do not lie on the same meridian.

83 (IIIB25). Strabo, Geography *2.1.23–6.*
(23) He [Eratosthenes] thus renders the second sealstone by the form of a rough outline, but he renders the third sealstone much more roughly, for several reasons. First, as already mentioned, the side from the Kaspian Gates to Karmania, common to the third and second sealstones, has not been defined distinctly, and then the Persian Gulf breaks into the southern side, as he himself says, and thus he was forced to take the line from Babylon as if it were straight, through Sousa and Persepolis to the borders of Karmania and Persis, on which he was able to find a measured route, being in total slightly more than 9,000 stadia. This he calls the southern side but he does not say that it is parallel to the northern. It is clear that the Euphrates, by which he marks off the western side, is nothing like a straight line, but after flowing from the mountains to the south it then turns toward the east and then back to the south until it empties into the sea. He makes clear that the river is not straight, in showing the shape of Mesopotamia, which is created by the confluence of the Tigris and the Euphrates, and resembles a rower's cushion, as he says. Moreover, for the portion from Thapsakos to Armenia he does not have a complete measurement like that of the western side which is marked off by the Euphrates, and he reports that he cannot say how much further the distance is to Armenia and the northern mountains, as it is unmeasured. Because of all this he says that he represents

the third portion very roughly. And he says that he collected the distances from many reports of those who had worked out the stopping points, some of which he says were without titles.

[Strabo's criticisms of Hipparchos omitted.]

(24) Thus he [Eratosthenes] says that he has shown the third portion roughly, with a length of 10,000 stadia from the Kaspian Gates to the Euphrates, and in dividing it into portions he set down the measurements as he found them already recorded, beginning in reverse from the Euphrates and its crossing at Thapsakos. As far as the Tigris, where Alexander crossed it, he writes 2,400 stadia, and then to the following places through Gaugamela, the Lykos, Arbela, and Ekbatana (by which Dareios fled from Gaugamela to the Kaspian Gates) he fills out with 10,000, having an excess of only 300 stadia. This is how he measures out the northern side, not having placed it parallel with the mountains or with the line through the Pillars, Athens, and Rhodes. For Thapsakos is far away from the mountains, and the mountains and the route from Thapsakos come together at the Kaspian Gates. These are the northern portions of the boundary.

(25) Having thus represented the northern side, he says that the southern cannot be taken along the sea because the Persian Gulf breaks into it, but from Babylon through Sousa and Persepolis to the boundaries of Persis and Karmania it is 9,200 stadia, which he calls the southern side, but he does not say that the southern is parallel to the northern. He says that the difference in length that occurs between the assumed northern and southern sides is because the Euphrates, having flowed to the south up to a point, turns more toward the east.

(26) Of the two flanking sides, he first speaks about the western. What it is like and whether it is one or two [lines] is considered uncertain. He says that from the Thapsakos crossing along the Euphrates to Babylon is 4,800 stadia, and from there to the outlet of the Euphrates and the city of Teredon is 3,000. But from Thapsakos to the north it has been measured as far as the Armenian Gates and is about 1,100, but through Gordyaiane and Armenia it is unknown and thus omitted. The eastern side, that which goes through Persis lengthwise from the Erythraian Sea somewhat toward Media and the north, he believes is no less than 8,000, and from certain promontories, over 9,000. The remainder through Paraitakene and Media to the Kaspian Gates is about 3,000. The Tigris and the Euphrates flow from Armenia to the south, and when they pass the mountains of Gordyaiane go around a great circle and enclose the large territory of Mesopotamia and then turn toward the winter sunrise and the south, especially the Euphrates. It constantly

becomes closer to the Tigris around the Wall of Semiramis and the village of Opis (from which it is only 200 stadia). Flowing through Babylon, it empties into the Persian Gulf. Thus it happens, he says, that the shape of Mesopotamia and Babylonia is like a cushion on a rowing bench. This is what Eratosthenes has said.

84 (IIIB26). Strabo, Geography 2.1.27.
He [Eratosthenes] says that the third section is bounded on its northern side by a line from the Kaspian Gates to the Euphrates, 10,000 stadia, but later he adds that the southern side, from Babylonia to the borders of Karmania, is slightly more than 9,000, and the western side from Thapsakos along the Euphrates to Babylon is 4,800 stadia, and to the outlet 3,000. As for the distance north of Thapsakos, one part has been measured at 1,100, but the remainder is unknown.

85 (IIIB27). Strabo, Geography 2.1.34.
He [Hipparchos] says that he [Eratosthenes] records the distance from Babylon to the Kaspian Gates as 6,700 stadia, and to the borders of Karmania and Persis over 9,000, following a line made straight to equinoctial east. This is perpendicular to the side common with the second and third sealstones, and thus a right-angle triangle is created with the right angle on the boundaries of Karmania and the hypotenuse shorter than one of the sides of the right angle, which necessarily puts Persis into the second sealstone. But I have said in regard to this that he [Eratosthenes] does not take [the distance] from Babylon to Karmania on a parallel, nor does he say that the straight line that separates the sealstones is a meridian, so he [Hipparchos] cannot speak against him. His further charge is also not good, since he [Eratosthenes] said that from the Kaspian Gates to Babylon was as already mentioned, and to Sousa 4,900 and from there to Babylon 3,400.

86 (IIIB34). Strabo, Geography 15.3.1.
According to Eratosthenes, the length of the territory toward the north and the Kaspian Gates is about 8,000, advancing by certain promontories, and the remainder to the Kaspian Gates is no more than 2,000. The width, in the interior, from Sousa to Persepolis, is 4,200 stadia, and from there to the border of Karmania an additional 1,600. The tribes living in the country are the so-called Pateischoreis, Achaimenidai, and the Magoi. They have chosen a certain holy life, but the Kyrtioi and Mardoi are piratical, and others are farmers.

87 (IIIB38, IIIB31). Strabo, Geography *16.1.21–2.*
(21) Mesopotamia is named from what it is. As I have said, it lies between the Euphrates and Tigris, and thus the Tigris washes only the
eastern side, and the Euphrates the western and southern. To the north
is the Tauros, which separates Armenia from Mesopotamia. The greatest distance that they are apart from one another is toward the mountains, which would be the same as Eratosthenes has said, 2,400, from
Thapsakos—where the ancient bridge over the Euphrates was—to the
crossing of the Tigris where Alexander himself crossed. The least is
slightly more than 200 somewhere around Seleukeia and Babylon. The
Tigris flows through the lake called Thopitis, through the middle of its
width, and crossing to the opposite edge sinks under the earth with a
great noise and upward blasts. It is invisible for a distance, and then
appears again not far from Gordyaia. It thus runs through it so vehemently, as Eratosthenes says, that although it is generally salty and
without fish, this part is sweet, with a strong current, and full of fish.
    (22) The contracting of Mesopotamia goes on for a great length,
somewhat like a boat, with the Euphrates making most of the circumference, and it is 4,800 stadia from Thapsakos as far as Babylon, as Eratosthenes says, and from Zeugma in Kommagene, where Mesopotamia
begins, to Thapsakos is no less than 2,000 stadia.

88. Strabo, Geography *14.2.29.*
The places lying in a straight line [from Tomisa on the Euphrates] as far
as India are the same in Artemidoros as Eratosthenes. But Polybios
says that the former must be the most trusted in regard to those places.
He begins from Samosata in Kommagene, which lies at the crossing and
at Zeugma, and says that to Samosata, from the boundaries of Kappadokia around Tomisa across the Tauros, is 450 stadia.

89 (IIIB32). Strabo, Geography *11.14.8.*
The Tigris rushes from the mountainous territory near the Niphates,
and the flow remains unmixed [with Lake Thopitis] because of its quick-
ness, from which comes its name, since the Medes call an arrow "tigris."
And while it has many types of fish, there is only one type in the lake.
Around the innermost part of the lake the river falls into a pit and
flows underground for some distance, coming up around Chalonitis.
From there it goes down toward Opis and the so-called Wall of Semiramis, leaving the Gordyaioi and all Mesopotamia on the right, but the
Euphrates, on the contrary, has the same territory on the left. Coming
near to one another and producing Mesopotamia, the former runs

through Seleukeia to the Persian Gulf and the latter through Babylon, which I said somewhere in my discussion against Eratosthenes and Hipparchos.

90 (IIIB37). Strabo, Geography *16.1.15.*
A large amount of asphalt is produced in Babylonia, about which Eratosthenes says that the liquid kind, which is called naphtha, is found in Sousis, but the dry kind, which can be solidified, is in Babylonia. There is a fountain of it near the Euphrates, and at the time of flooding by snow melt it fills and overflows into the river. Large lumps are formed that are suitable for structures of baked brick.

91 (IIIB35). Stephanos of Byzantion, "Assyria."
Assyres is said by Eratosthenes, as well as Illyres, from the Illyrians.

The Fourth Sealstone (Arabia, Egypt, and Aithiopia)

92 (IIIA29). Strabo, Geography *2.1.32.*
The fourth sealstone is composed of Arabia Eudaimon, the Arabian Gulf, all Egypt, and Aithiopia. The length of this portion will be that bounded by the two meridians, one of which is drawn through its western point and the other through the most eastern. The width will be between two parallels, one of which is drawn through the most northern point and the other [through] the most southern.

93 (IIIB41). Pliny, Natural History *6.108.*
It [the Erythraian Sea] is divided into two gulfs. The one in the east is called the Persian, 2,500 miles around, as Eratosthenes reports. Opposite is Arabia, which is 1,500 miles around, and on the other side is the second bay, called the Arabian, and the ocean that flows in is the Azanian. At its entrance the width of the Persian [Gulf] is, according to some, five miles, and to others, four miles. From there to the inner gulf has been determined to be nearly 1,125 miles in a straight line, and its form is the shape of a human head.

94 (IIIB39). Strabo, Geography *16.3.2–6.*
(2) The Persian Gulf is also called the Persian Sea. Eratosthenes says the following about it: he says that its mouth is so narrow that from Harmozai, the promontory of Karmania, one can look across to that of Makai in Arabia. From its mouth the right-hand coast, being curved, is

at first turned slightly to the east from Karmania, and then bends toward the north, and afterward toward the west as far as Teredon and the mouth of the Euphrates, consisting of the Karmanian coast, and part of the Persian, Sousian, and Babylonian, about 10,000 stadia, concerning which I have already spoken. From there to its mouth is the same distance. This, he says, is according to Androsthenes the Thasian, who sailed with Nearchos and also by himself. It is thus clear that this sea is only slightly smaller than the Euxeinos Sea. He [Eratosthenes] says that he [Androsthenes], who sailed around the gulf with an expedition, said that beyond Teredon—having the continent on his right and sailing along the coast—is the island of Ikaros, with a temple on it sacred to Apollo and an oracle of Tauropolos.

(3) After sailing along Arabia for 2,400 stadia there is the city of Gerrha, lying on a deep gulf, where Chaldaians exiled from Babylon live. The land is salty and they have houses of salt, and since flakes of salt come away because of the heat of the sun and fall off, they sprinkle the walls with water and keep them solid. The city is 200 stadia from the sea. The Gerrhaians mostly trade by land for Arabian goods and aromatics. In contrast Aristoboulos says that the Gerrhaians generally travel on rafts to Babylonia for trade, sailing up the Euphrates with their goods to Thapsakos, and distributing them everywhere from there by land.

(4) Sailing farther, there are other islands, Tyros and Arados, which have temples like those of the Phoenicians. Those living there say that the islands and cities that have the same names are Phoenician settlements. These islands are ten days' sail from Teredon and one day from the mouth at the promontory of Makai.

(5) Both Nearchos and Orthagoras say that the island of Ogyris lies on the Southern Ocean at a distance of 2,000 stadia, and on it the grave of Erythras can be seen, a large mound planted with wild palms. He was king of that region and left the sea named after himself. They say that these things were pointed out to them by Mithropastes son of Arsites the satrap of Phrygia, who had been exiled by Dareios and passed his time on the island, joining them when they landed in the Persian Gulf and attempting through them to return home.

(6) Along the entire coast of the Erythraian Sea are trees in the depths that are like laurel and olive, completely visible at low tide but completely covered at high tide, although the land lying above is without trees, thus intensifying the peculiarity. Concerning the region of the Persian Sea, which, as I have said, forms the eastern side of Arabia Eudaimon, this is what Eratosthenes has said.

95 (IIIB48). Strabo, Geography *16.4.2–4.*

(2) But I return to Eratosthenes, who sets forth what he knows about Arabia. He says, concerning its northerly or desert part, which is between Arabia Eudaimon and Koile Syria and Judaea as far as the recesses of the Arabian Gulf (from Heroonpolis, which is at the recess of the Arabian Gulf near the Nile), that it is 5,600 stadia toward Nabataean Petra to Babylon, entirely toward the summer sunrise, through the adjacent Arabian tribes, the Nabataeans, Chaulotaioi, and Agraioi. Beyond these is Eudaimon, which extends for 12,000 stadia toward the south as far as the Atlantic Ocean. The first people there, beyond the Syrians and Judaeans, are farmers. Beyond these the land is very sandy and wretched, with a few palm trees, a thorny plant, and the tamarisk, and water by digging, just as in Gedrosia. The tent-dwelling Arabians and camel herders are there. The extremities, toward the south, rising opposite to Aithiopia, are watered by summer rains and have two sowings, like India, and the rivers are consumed by plains and lakes. It is fertile and abundant in honey production. It has plenty of fatted animals (except for horses, mules, and pigs), and all kinds of birds except geese and chickens. The four most numerous peoples living in the extremity of the previously mentioned territory are the Minaioi, in the district toward the Erythraian Sea, whose largest city is Karna or Karnana; next to these are the Sabaioi, whose metropolis is Mariaba; third are the Kattabaneis, extending to the straits and the crossing of the Arabian Gulf, whose royal seat is called Tamna; and, farthest toward the east are the Chatramotitai, whose city is Sabata.

(3) All are monarchies and prosperous, beautifully furnished with temples and palaces. Their houses are like those of the Egyptians regarding the joining of the beams. The four districts have more territory than the Egyptian Delta. No child succeeds to the kingship of his father, but the child of a distinguished person who was born first after the accession of the king. At the same time that someone accedes to the throne, the pregnant wives of distinguished men are recorded and guards are placed. The son who is born first to one of them is by law adopted and raised in a royal fashion to be the successor.

(4) Kattabania produces frankincense, and Chatramotitis myrrh, and these and other aromatics are traded with merchants. They come there from Ailana, arriving in Minaia in 70 days. Ailana is a city on the other recess of the Arabian gulf, the one opposite Gaza called the Ailanitic, as I have already said. The Gerrhaioi, however, arrive at Chatramotitis in 40 days. The part of the Arabian Gulf along the side of Arabia, beginning at the Ailanitic recess, was recorded by those around Alexander,

especially Anaxikrates, as 14,000 stadia, although this is said to be too much. The part opposite the Trogodytic territory, which is on the right when sailing from Heroonpolis, as far as Ptolemais and the elephant-hunting territory, is 9,000 stadia to the south and slightly toward the east, and then, as far as the straits, 4,500 somewhat more to the east. The straits are created by a promontory called Deire, and a small town with the same name, in which the Ichthyophagoi live. Here, it is said, there is a pillar of Sesostris the Egyptian that records his crossing in hieroglyphics. It appears that he was the first to subdue the Aithiopic and Trogodytic territory, and then he crossed into Arabia and proceeded against all Asia. Because of this, fortifications of Sesostris are identified everywhere, as well as reproductions of temples to the Egyptian gods. The straits at Deire narrow to 60 stadia. But these are not called the straits now, but, sailing further along, where the passage between the continents is about 200 stadia, there are six islands that come in succession and fill up the crossing, leaving extremely narrow passages through which boats carry goods across, and these are called the straits. After the islands, the next sailing, following the bays along the myrrh-bearing territory toward the south, and east as far as the kinnamomon-bearing territory, is about 5,000 stadia. It is said that until now no one has gone beyond this region. There are not many cities on the coast, but many beautiful settlements in the interior. This, then, is the account of Eratosthenes about Arabia.

96 (IIIB36). Strabo, Geography *16.1.12.*
Eratosthenes, mentioning the lakes near Arabia, says that when the water is unable to exit, it opens underground passages and flows underground as far as Koile Syria, pressing into the region of Rhinokoloura and Mt. Kasion, creating lakes and pits there.

97 (IIIB50). Pliny, Natural History *6.163.*
Timosthenes figured the entire gulf at four [?] days' sail in length and two in width, and 7½ miles at its narrowest, Eratosthenes 1,200 miles from the mouth on either side, Artemidoros 1,750 miles for the length of the Arabian side and on the Trogodytic side 1184½ miles as far as Ptolemais.

98 (IIIB51). Strabo, Geography *17.1.1–2.*
(1) In making the rounds of Arabia we have included the gulfs that tighten it up and make it a peninsula, the Persian and Arabian, and at the same time have described parts of Egypt and Aithiopia, that of the Trogodytai and those beyond them as far as the Kinnamomophoroi. The remaining [territory] that touches these peoples must be set forth,

the regions around the Nile. Afterward we will go across Libya, which is the last topic of the entire *Geography*, and also the assertions of Eratosthenes must be expounded.

(2) He says that the Nile is 900 or 1,000 stadia west of the Arabian Gulf, similar to the shape of a backward letter N, for, he says, after flowing from Meroë toward the north for about 2,700 stadia, it turns back toward the south and the winter sunset for about 3,700 stadia, and after coming almost opposite to the location of Meroë and projecting far into Libya, it makes the second turn and is carried to the north, 5,300 stadia, to the great cataract, turning aside slightly toward the east, and then 1,200 to the smaller one at Syene, and then 5,300 more to the sea. There are two rivers that empty into it, which come from certain lakes to the east and encircle Meroë, a good-sized island. One of these is called the Astaboras, flowing on the eastern side, and the other the Astapous, although some call it the Astasobas, saying that another is the Astapous, flowing from certain lakes to the south, and that this one makes almost the entire straight body of the Nile, created by filling with summer rains. Seven hundred stadia above the confluence of the Astaboras and the Nile is Meroë, a city with the same name as the island. There is another island above Meroë that is held by the Egyptian fugitives who revolted at the time of Psammitichos, called the Sembritai, which means "foreigners." They are ruled by a woman but are subject to those in Meroë. In the lower districts on either side of Meroë, along the Nile toward the Erythraian Sea, live the Megabaroi and Blemmyes, subject to the Aithiopians and bordering the Egyptians, and along the sea are the Trogodytes. The Trogodytes opposite Meroë lie ten or 12 days' journey from the Nile. On the left side of the course of the Nile in Libya live the Noubai, a large group of people who begin at Meroë and extend as far as the bends, not subject to the Aithiopians but divided into a number of separate kingdoms. The extent of Egypt along the sea from the Pelousiac to the Kanobic mouth is 1,300 stadia. Eratosthenes has these things.

99 (IIIB52). Proklos, Commentary on Plato's Timaios *p. 37b.*
Others say that the increase in the Nile is because of the rain pouring into it, which is said explicitly by Eratosthenes.

Libya

100 (IIIB59). Strabo, Geography *17.3.1–2.*
(1) I will next speak about Libya, which is the remaining part of the entire *Geography*. I have previously said much about it, but now additional

appropriate matters must also be mentioned, adding what has not been previously said. Those who have divided the inhabited world have divided it unequally, but Libya is so much lacking in being a third part that even if it were combined with Europe it would not seem to be equal to Asia. Perhaps it is smaller than Europe, and greatly so in regard to its importance, for much of the interior and coast along the Ocean is desert, dotted with small settlements that are for the most part scattered and nomadic. In addition to the desert, it abounds in wild beasts that drive [the inhabitants] away even from areas capable of habitation, and it occupies much of the burned zone. However, all of the coast opposite us is prosperously settled—that between the Nile and the Pillars—especially the part that is under the Karchedonians, although even here portions are found without water, such as those around the Syrtes, Marmaridai, and Katabathmos. It is in the shape of a right-angled triangle, as conceived on a level surface, having as its base the seacoast opposite us from Egypt and the Nile as far as Maurousia and the Pillars, as the side perpendicular to this that which is formed by the Nile as far as Aithiopia and extended by us to the Ocean, and as the hypotenuse to the right angle the entire coast of the Ocean between the Aithiopians and the Maurousians. That at the extremity of the previously mentioned triangle, lying somewhat within the burned [zone], we can speak about only from conjecture because it is inaccessible, so we cannot speak of the greatest width of the land, although we have said in a previous section this much, that going to the south from Alexandria to Meroë, the royal capital of the Aithiopians, is about 10,000 stadia, and from there in a straight line to the boundaries between the burned and inhabited regions is another 3,000. Then this should be put down as the greatest width of Libya, 13,000 or 14,000 stadia, with the length slightly less than double. This is then the totality about Libya, but I must speak about each [region], beginning from the western parts, the most famous.

(2) At this point is the strait called the Pillars of Herakles, of which I have spoken often. Going outside the strait at the Pillars, having Libya on the left, there is the mountain that the Hellenes call Atlas and the barbarians Dyris. From these something projects farthest toward the west of Maurousia, called the Koteis. Nearby is a small town by the sea that the barbarians call Tinx, but Artemidoros calls it Lynx and Eratosthenes Lixos. It lies across the strait from the Gadeirenes, 800 stadia across the sea, which is about as far as each lies from the strait at the Pillars. To the south of Lixos and the Koteis lies the gulf called Emporikos, having settlements of Phoenician merchants. The entire seacoast encompassed by this gulf is indented, but the gulfs and projections

should be removed, according to the triangle shape I have outlined. One must conceive that the continent increases to the south and east. The mountain that extends through Maurousia from the Koteis as far as the Syrtes is inhabited, both it and the others that are parallel, first by Maurousians but deep within the territory by the most numerous of the Libyan peoples, called the Gaitoulai.

101 (IIIB53). Pliny, Natural History *5.39.*
Eratosthenes records the land route from Kyrene to Alexandria as 525 miles.

102 (IIIB54). Strabo, Geography *2.1.40 (part of F65).*
[Eratosthenes says] that it is over 13,000 [stadia] from Alexandria to Karchedon.

103 (IIC20). Pliny, Natural History *5.40.*
Polybios and Eratosthenes, who are considered most diligent, made it 1,100 miles from the Ocean to Great Carthage, and 1,628 miles from there to the Kanopos, the nearest mouth of the Nile.

104 (IIIB56). Strabo, Geography *2.5.20.*
Of the Syrtes, the lesser one is about 1,600 stadia in circumference, and the islands Meninx and Kerkina lie on either side of its mouth. Eratosthenes says that the Great Syrtes has a circuit of 5,000 and it is 1,800 deep, from the Hesperides to Automala and the boundary in that region between the Kyrenaika and the rest of Libya.

105 (IIIB57). Pliny, Natural History *5.41.*
These seas do not include many islands. The most famous is Meninx, 25 miles across and 22 wide, which Eratosthenes called Lotophagitis. It has two towns, Meninx on the African side and Thoas on the other, [the island] itself located 1½ miles from the promontory on the right side of the Lesser Syrtis. One hundred miles from it off the left-hand side is Kerkina, with the free city of the same name. It is 25 miles long and half that across where it is widest, but no more than five at the end. Joined to it by a bridge is tiny Kerkinitis, toward Carthage. About 50 miles from these is Lopadousa, six miles long, then Gaulos and Galata, whose earth kills scorpions, the terrible animal of Africa.

106 (IIIB58). Strabo, Geography *3.5.5.*
For this reason [a mythological explanation] some believe that the peaks at the strait are the Pillars, others at Gadeira, and others that they lie

farther outside of Gadeira. Some assume that the Pillars are Kalpe and Abilyx, which is the mountain opposite in Libya and which Eratosthenes says is situated in Metagonion, a Nomadic people, and others that they are the islets near each [mountain], one of which is named Hera's Island. Artemidoros speaks of Hera's Island and her temple, and he mentions another, but neither Mount Abilyx or the Metagonian people. Some transfer the Planktai and Symplegades here, believing these to be the Pillars that Pindar calls the Gadeiran Gates, saying that they were the farthest point reached by Herakles. And Dikaiarchos, Eratosthenes, and Polybios and most of the Hellenes believe that the Pillars are around the narrows.

107 (IIIB60). Strabo, Geography *17.3.8.*
Artemidoros disagrees with Eratosthenes because he says that a certain city near the western extremities of Maurousia is Lixos rather than Lynx, and that there are a large number of Phoenician cities that have been destroyed and of which there are no traces to be seen, and because he says that the air among the western Aithiopians is brackish, and that the air at the hours of daybreak and the afternoon is thick and misty.

The Northeastern Part of the Inhabited World

108 (IIIB20, IIIB63). Strabo, Geography *11.8.8–9.*
(8) Eratosthenes says that the Arachotoi and Massagetai are alongside the Baktrians to the west along the Oxos, and that the Sakai and Sogdianoi and all their territory lie opposite India, although the Baktrians only for a small distance, for they are mostly along the Paropamisos. The Sakai and Sogdians are separated by the Iaxartes, and the Sogdianoi and Baktrians by the Oxos, and the Tapyroi live between the Hyrkanians and the Arioi. In a circuit around the [Kaspian] sea, after the Hyrkanians are the Amardoi, Anariakai, Kadousioi, Albanoi, Kaspioi, and Ouitioi, and perhaps others, until the Skythians are reached. On the other side of the Hyrkanians are the Derbikes, and the Kadousioi touch the Medes and the Matianoi below the Parachoathras.

(9) He says that these are the distances: from Kaspios to the Kyros is about 1,800 stadia, and then to the Kaspian Gates 5,600, to Alexandria among the Arioi, 6,400, then to the city of Baktra, also called Zariaspa, 3,800, then to the Iaxartes River, to which Alexander came, about 5,000, a total of 22,670. He [Eratosthenes] also says that the distances from the Kaspian Gates to India are as follows: to Hekatompylos they

say is 1,960, to Alexandria among the Arioi 4,530, then to Prophthasia in Drange 1,600 (others say 1,500), then to the city of the Arachotoi 4,120, then to Ortospana and the meeting of three roads from Baktra 2,000, and then to the borders of India 1,000, a total of 15,300. It must be believed that the length of India is a distance in a straight line, that from the Indos as far as the eastern sea.

109 (IIIB67). Strabo, Geography 11.7.3.

The rivers flowing through Hyrkania are the Oxos and Ochos, which empty into the sea. Of these, the Ochos also flows through Hesaia, but some say that the Ochos empties into the Oxos. Aristoboulos says that the Oxos is the largest he had seen in Asia, except those in India. He also says that it is navigable (he and Eratosthenes took this from Patrokles), and that many Indian goods come down it to the Hyrkanian Sea, and from there are carried over to Albania by means of the Kyros River, brought down through the successive places to the Euxeinos.

110 (IIIB68). Strabo, Geography 11.6.1.

The second portion begins from the Kaspian Sea, where the first comes to an end. The same sea is also called the Hyrkanian. It is necessary to speak first about the sea and the peoples living around it. It is the gulf that extends from the Ocean to the south, somewhat narrow at its entrance but becoming wider as it goes inland, especially around its recess, where it is about 5,000 stadia. Sailing from the entrance to the recess would be slightly more, since it nearly touches the uninhabited region. Eratosthenes says that the circuit of this sea was well known to the Hellenes, and the portion along the Albanians and Kadousioi is 5,400 stadia, and the portion along the Anariakoi, Mardoi, and Hyrkanoi to the mouth of the Oxos River is 4,800, and from there to the Iaxartes, 2,400.

111 (IIIB71). Pliny, Natural History 6.36.

Eratosthenes makes the measurement on the southeast along the coast of the Kadousiai and Albania as 9,000 stadia, and from there through the Atiakoi, Amarboi, and Hyrkanoi to the mouth of the Zonos River 4,900 stadia, and from there to the mouth of the Iaxartes 2,400, which makes a total of 1,575 miles.

112 (IIIB72). Scholia to Apollonios of Rhodes 2.1247.

Eratosthenes says that those called the Kaukasiai are near the Kaspian Sea.

113 (IIIB73). Strabo, Geography *11.2.15.*
Eratosthenes says that the Kaukasos is called the Kaspian by those living there, perhaps derived from the Kaspioi.

114 (IIIB79). Ammianus Marcellius 22.8.10.
The sail around its entire shore [of the Euxine], as if the circuit of an island, is 23,000 stadia, as Eratosthenes asserts, as well as Hekataios, Ptolemy, and others who have investigated such issues very carefully, and all geographers agree that it is shaped like a drawn Skythian bow.

115 (IIIB78). Pliny, Natural History *6.3.*
The dimensions of the Pontos from the Bosporos to the Maiotic Lake some make as 1,438½ miles, but Eratosthenes 100 less.

116 (IIIB77). Pliny, Natural History *5.47.*
The distance from the mouth of the Pontos to the mouth of the Maiotis Eratosthenes records as 1,545 miles.

Anatolia

117 (IIIB80). Scholia to Euripides, Medea *2.*
Eratosthenes in his geographical treatise says that the passage [at the Symplegades] is narrow and crooked, because of which it was imagined that sailors would be caught in the rocks.

118 (IIIB82). Tzetzes on Lykophron, Alexandra *1285.*
Eratosthenes calls them the Synormades. He says that they are hidden and unseen [rocks] around the Euxeinos Sea, especially in the narrows.

119 (IIIB84). Strabo, Geography *11.14.7.*
There are a number of rivers in the territory, the best known of which are the Phasis and the Lykos, which empty into the Pontic Sea. Eratosthenes has wrongly put down the Thermodon instead of the Lykos.

120 (IIIB75). Scholia to Apollonios of Rhodes 2.399.
The Phasis flows from the mountains of Armenia, as Eratosthenes says. It empties into the sea at Kolchis.

121 (IIIB76). Scholia to Apollonios of Rhodes 4.131.
The Titenis River, from which the land of Titenis is named, is mentioned by Eratosthenes in his *Geography*.

122 (IIIB86). Stephanos of Byzantion, "Gangra."
Eratosthenes has "Gangra" in the neuter.

123 (IIIB87). Stephanos of Byzantion, "Amaxa."
Amaxa, a district of Bithynia, according to Eratosthenes.

124 (IIIB88). Stephanos of Byzantion, "Tarsos."
Eratosthenes has Tersos . . . and Eratosthenes also says Tarsenoi.

125 (IIIB89). Eustathios, Commentary on Dionysios Periegetis 867.
Eratosthenes says that the name of the city is so called by those there
from Zeus Tersios.

126 (IIIB85). Pliny, Natural History *5.127.*
Eratosthenes records people from Asia who have died out: the Solymoi,
Leleges, Bebrykes, Kolykantioi, and Tripsedoi.

The Mediterranean

127 (IIIB92). Pliny, Natural History *3.75.*
Eratosthenes, however, calls whatever is between the mouth of the sea
and Sardinia the Sardoan, from there to Sicily the Tyrrhenian, from
there to Crete the Sikelian, and from there the Cretan.

128 (IIB28). Strabo, Geography *2.5.24.*
The sea passage from Rhodes to Alexandria, with a north wind, is about
4,000 stadia, but along the coast it is double that. Eratosthenes says
that this is the estimate of sailors regarding the crossing of the sea,
some saying this, and others not shrinking from saying 5,000, but that
he, using the shadow of a gnomon, found it to be 3,750.

129 (IIIB94). Strabo, Geography *10.4.5.*
Eratosthenes says that from the Kyrenaika to Krioumetopon is 2,000,
and less from there to the Peloponnesos. . . .

130 (IIIB91). Strabo, Geography *14.6.4–5.*
(4). . . comparing what Damastes says, who gives the length of the island [Cyprus] as running from north to south, from Hierokepias, as he
says, to Kleides. Nor is Eratosthenes correct, although complaining
about him [Damastes], when he says that Hierokepias is not in the
north, but the south. It is not in the south, but the west, as it lies on the

western side, where Paphos and Akamas are. Such is the location of the
position of Cyprus.

(5) It is not inferior to any other island in prosperity, for it is good in
wine and good in olives, and sufficient in grain for its use. There are
abundant copper mines at Tamassos, in which *chalkanthes* is produced,
and also verdigris of copper, useful for its medicinal purposes. Eratosthenes says that formerly the plains were overrun with woods, and thus
there were thickets spread over them that were uncultivated. Mining
helped this a little, since the trees would be cut down to burn the copper
and silver, and in addition there was shipbuilding, since sailing on the
sea was safe through sea power. Yet because it was not possible to prevail, those wishing or able were allowed to cut them down and to possess the cleared land as private property, tax free.

Europe

131 (IIIB66, IIIB96). Strabo, Geography *2.1.41.*
But let it now be said that Timosthenes, Eratosthenes, and those even
earlier were completely ignorant of Iberika and Keltika, and immensely
more so about Germanika and Brettanika, as well as the territory of the
Getai and the Bastarnai. They also happened to be somewhat ignorant
of those in the region of Italy, Adria, the Pontos, and those portions beyond to the north, although such [statements] are perhaps finding fault.
In regard to remote areas, Eratosthenes says that he records the distances that have been handed down, but does not validate them, reporting them as they have been received, although at times adding "by means
of a more or less straight line." One cannot put to a strict test those distances that do not agree with one another. But this is what Hipparchos
attempts to do, both in the examples formerly mentioned, and where he
establishes the distances around Hyrkania as far as Baktria and to the
people living beyond, as well as from Kolchis to the Hyrkanian Sea.

132 (IIIB96). Strabo, Geography *2.4.2 (part of F14).*
I have said that Eratosthenes was ignorant of the western and northern
parts of Europe. But there must be leniency toward him and Dikaiarchos, as they had not seen those places.

133 (IIIB110, IIIB123, IIIB119). Strabo, Geography *2.4.4.*
Next, he [Polybios] corrects the statements of Eratosthenes, some correctly, but in others what he says is worse. For example, when he [Era-
tosthenes] says that from Ithaka to Korkyra is 300 [stadia], he [Polybios]

says that it is more than 900; [Eratosthenes] having set down 900 from Epidamnos to Thessalonike, he [Polybios] says it is more than 2,000, but in these he [Eratosthenes] is correct. Since he [Eratosthenes] says that it is 7,000 from Massalia to the Pillars, which are 6,000 from the Pyrenes, he [Polybios] himself speaks in error in saying that it more than 9,000 from Massalia and from the Pyrenes a little less than 8,000, for he [Eratosthenes] is nearer to the truth. Those today agree that if one cuts through the irregularities of the roads, the length of the whole of Iberia, from the Pyrenes to its western side, is no more than 6,000 stadia. But he [Polybios] puts the Tagos River at 8,000 in length from its source as far as its mouth, without any of its bends (this is not being geographical), but speaking of a straight line, although the sources of the Tagos are more than 1,000 stadia from the Pyrenes. On the other hand, he is correct when he proclaims that Eratosthenes is ignorant of Iberika, because he proclaims things that conflict with one another, such as when he says that the exterior as far as Gadeira is inhabited by Galatai—if they do possess the western part of Europe as far as Gadeira—he forgets this and nowhere records the Galatai in his circuit of Iberia.

134 (IIIB97). Strabo, Geography *2.1.40.*
Then he [Hipparchos] sets forth what Eratosthenes has said about the places after the Pontos, that he said there are three promontories coming down from the north: one, on which is the Peloponnesos, the second, the Italian, and the third, the Ligystikan, which cuts off the Adriatic and Tyrrhenian Gulfs. Having generally expounded this, he [Hipparchos] attempts to test each statement about them by geometry rather than geography. But the errors made about them by Eratosthenes are so numerous, as well as those by Timosthenes, the writer of *On Harbors*, whom he praises more than the others, although he refutes him, disagreeing on most things, that I do not believe it worthwhile to arbitrate between them, or in regard to Hipparchos, since they are entirely in error. Even he [Hipparchos] passes over their mistakes, not correcting them but only proving that they were false or contradictory. Perhaps one could accuse him [Eratosthenes] of this because he says that there are three promontories of Europe, putting down that which is the Peloponnesos as one, although it is split into many parts. For Sounion makes a promontory, just like Lakonia, which is not much less to the south than Maleia and includes a notable gulf. And from the Thracian Chersonesos up to Sounion the Melanian Gulf and those as far as the Makedonian are included. Even if we were to overlook this, most of the distances are obviously wrong and prove that his ignorance of these places

is excessive, without needing geometrical proofs but only those that are obvious and can be immediately witnessed.

135 (IIIB97). Strabo, Geography *2.4.8.*
Since Europe extends into a number of promontories, he [Polybios] describes it better than Eratosthenes, although not adequately. He spoke only of three, the one going down to the Pillars, on which is Iberia, the one to Porthmos, on which is Italy, and the third, down to Maleia, on which are all the peoples between the Adria and the Euxeinos and the Tanais.

136 (IIIB100). Strabo, Geography *1.2.20 (part of F11).*
Thrace, where it touches Makedonia, happens to make a turn to the south and forms a promontory into the open sea. . . . Eratosthenes did not perceive this, although he conjectured it. Nevertheless he told about the turn of the land that I have mentioned.

137 (IIIB101). Stephanos of Byzantion, "Ichnai."
Ichnai, a Makedonian city . . . Eratosthenes says it is the same as Achne.

138 (IIIB102). Life of Aratos *I.*
The citizens in Attika are called Athenaioi. Those in Euboia are called Athenetai, as Eratosthenes says in the first [third?] book of his *Geography*.

139 (IIIB103). Strabo, Geography *8.7.2.*
Helike was submerged two years before Leuktra. Eratosthenes says that he himself saw the place, and the ferrymen say that a bronze Poseidon stood upright in the strait, with a hippocamp in his hand, a danger to those fishing with nets.

140 (IIIB105). Strabo, Geography *8.8.4.*
Eratosthenes says that around Pheneos the river called the Anias makes a lake in front of the city and flows down into strainers which are called "zerethra." When these are stopped up, the water at times overflows into the plains, but when they are opened up again, it comes out of the plains all together and empties into the Ladon and the Alpheus, so that once at Olympia the land around the temple was flooded and the lake was diminished. The Erasinos, flowing around Stymphalos, goes under the mountain and reappears again in the Argive territory. Because of this Iphikrates, when he was besieging Stymphalos and accomplishing

nothing, attempted to block up the sinkhole with many sponges that he provided, but ceased because of a sign from Zeus."

141 (IIIB106). Eustathios, Commentary on Iliad *2.612.*
... that Eratosthenes says that the *phellos* exists in Arkadia, among the *prinoi* and *dryoi*, some of which are called the *thelyprinos*.

142 (IIIB108). Stephanos of Byzantion, "Agraioi."
There are other peoples near Akarnania ... called the Agraeis, as by Eratosthenes.

143 (IIIB109). Stephanos of Byzantion, "Dyrrhachion."
Eratosthenes in the third book of his *Geography*: the Taulantioi live there. The Hellenic city of Epidamnos in the Chersonesos is called Dyrrhachion. Its rivers are the Drilon and Aoos, around which are shown the graves of Kadmos and Harmonia.

144 (IIIB111). Stephanos of Byzantion, "Autariatai."
Autariatai, a Thesprotikan people: Charax ... and Eratosthenes.

145 (IIIB112). Scholia to Apollonios of Rhodes 4.1215.
Skylax says that the Nestaioi are an Illyrian people. Eratosthenes in the third book of his *Geography* says, "Among the Illyrians are the Nestaioi, among whom is the island of Pharos where the Phariai live."

146 (IIIB113). Periplous Dedicated to King Nikomedes 405–12.
Beyond is the great peninsula of Hyllis, which is about equal in size to the Peloponnesos. They say that 15 cities are on it, established by the Hylloi, being of the Hellenic race. They report that Hyllos the Heraklid was the founder, but that in time they became barbarians because of the neighboring peoples, as Timaios and Eratosthenes recorded.

147 (IIIB117). Stephanos of Byzantion, "Tauriskoi."
Tauriskoi, a people around Mt. Alpeion.... Eratosthenes says they are the Teriskoi, using the ε.

148 (IIIB98). Scholia to Apollonios of Rhodes 4.310.
Eratosthenes says in the third book of his *Geographika* that the two mouths of the Istros empty into the sea around the triangular island of Peuke. It is similar to the island of Rhodes, and is called Peuke because of its many pines.

149 (IIIB99). Scholia to Apollonios of Rhodes 4.284.
Eratosthenes in the third book of his *Geographika* [says] that it [the Is-tros] flows from uninhabited territory and around the island of Peuke.

150 (IIIB118). Caesar, Gallic War 6.24.
Thus the most fertile parts of Germania are around the Hercynian woods, which I see were known by report to Eratosthenes and certain Greeks, who called them the Orkynian.

151 (IIIB116). Strabo, Geography 5.2.6.
Neither, then, is Eratosthenes correct when he says that neither Kyrnos nor Sardon can be seen from the mainland.

152 (IIIB120). Strabo, Geography 3.4.7.
Between the turning aside of the Iberos and the heights of the Pyrenes, on which are set up the dedications of Pompeius, the first city is Tarra-kon, which has no harbor but is set on a bay and is equipped sufficiently with everything, and is now no less populated than Karchedon. It is naturally suited for the residence of the commanders and is the me-tropolis, so to speak, not only of the territory within the Iberos but much of that beyond. The Gymnesian Islands lie nearby, and Ebysos, notable islands, which suggests that the position of the city is propitious. Era-tosthenes says that it also has an anchorage.

153 (IIIB122). Strabo, Geography 3.2.11.
Eratosthenes says that the territory adjoining Kalpe is called Tartessis, and that Erytheia is the Island of the Blessed. Artemidoros contradicts him and says that this is a falsehood by him, just like that the distance from Gadeira to the Sacred Promontory is a five-day sail (although no more than 1,700 stadia), that the tides terminate at that point (although they exist around the circuit of the entire inhabited world), that the northerly part of Iberia is an easier means of access to Keltika than sailing by the Ocean, and everything else that he said while relying on Pytheas, because of his false pretensions.

Conclusion

154 (IB9). Strabo, Geography 17.1.19.
Eratosthenes says that banishment of foreigners is common among all barbarians, but the Egyptians are censured for this because of the tales concerning Bousiris in the Bousiritic Nome. Those later wish to accuse

this place of inhospitality, but, by Zeus, there was no king or *tyrannos* named Bousiris. Moreover, this is quoted as a reproach:

> To go to Egypt, a long and painful journey
> *[Odyssey 4.483].*

The lack of harbors adds greatly to the problem, and also that one could not enter the single existing harbor, the one at Pharos, because it was guarded by shepherds who were pirates, and who would attack those coming to anchor. The Karchedonians would drown any foreigner who sailed past them to Sardo or the Pillars. Because of this most things about the west are not believed. Moreover, the Persians would deceitfully guide ambassadors over circuitous roads and through difficult places.

155 (IIC24). Strabo, Geography 1.4.9.

Near the end of his treatise he [Eratosthenes] will not praise those who separate all the number of humanity into two groups, Hellenes and barbarians, as well as those who advised Alexander to consider the Hellenes as friends but the barbarians as enemies. He says that it is better to make such a distinction between good and bad characteristics, for there are many bad Hellenes or urbane barbarians, such as the Indians or Arians, or, moreover, the Romans and Karchedonians, who administer their governments so marvelously. Because of this, Alexander ignored his advisors, embraced as many distinguished men as possible, and showed them kindness.

SUMMARIES AND COMMENTARIES

Commentary to Book 1

F1. *Summary*: **Geography is a scholarly discipline. Men such as Homer and many others were the first to investigate it.**

Commentary: The opening of Strabo's *Geography* immediately shows his major debt to Eratosthenes, the first primary source mentioned. In fact Strabo's comments about the validity of geography as a scholarly discipline may derive from Eratosthenes' own opening, since he would stress the legitimacy of his new discipline. Strabo (and probably Eratosthenes) noted that geography was the concern of the φιλόσοφος, a word not to be translated with the misleading "philosopher" but better as "scholar" (or, more cumbersomely, "educated person"), a meaning in use since the fifth century BC (first generically by Plato, *Republic* 5.19, and then specifically about Euripides [Athenaios 13.561a] and Aristotle [Plutarch, *Letter to Apollonios* 27]). This is essentially the first extant use of the word "geography" and its relatives, although Strabo's dependence on Eratosthenes shows that it was by no means a new word. Slightly earlier than Strabo is a citation by Philodemos (*On Poetry* 5), in a list of disciplines contrasted with poetry (sentiments that may owe their origin to Eratosthenes and his comments on Homer [F2–11]). Philodemos came to Rome in the 70s BC and survived into the 30s; his acquaintance Cicero was the first to use "geography" in Latin (*Letters to Atticus* #24, 26, 27, all from 59 BC; Letter 26 uses the Greek).

Eratosthenes began his treatise by crediting the pioneers of geographical research. Presumably he was using the new word "geography" loosely to apply to his distinguished predecessors, avoiding the obvious semantic trap of how those who lived before the establishment of the discipline could have been geographers. Only Eratosthenes' discussion of Homer survives in any detail (F2–11), because of Strabo's disdain for Eratosthenes' views and his lengthy repudiation of them (E. Hönigmann, "Strabon von Amaseia" [#3], *RE* 2nd ser. 4 [1931] 144–7). In the extant fragments, Anaximandros and Hekataios are mentioned only once again (F12). That the second part of the list of early geographers also comes from Eratosthenes is probable but not certain, given Strabo's construction of the sentence. Demokritos of Abdera wrote a *Kosmographia* about which nothing is known: it is merely recorded as a title in the list of his works by Thrasyllos of Alexandria, the astrologer to the emperor Tiberius (Diogenes Laertios 9.46: the *Geographia* in this list is probably an anachronism). Eratosthenes' only other extant mention of Demokritos (F33) has no geographical context (but see F32). Eudoxos of Knidos was an important source for Eratosthenes, having created the concept of latitudinal zones, the κλίματα (supra, p. 6). Dikaiarchos of Messana wrote a *Circuit of the Earth* (Paul T. Keyser, "The Geographical Work of Dikaiarchos," in *Dicaearchus of Messana* [ed. Fortenbaugh and Schütrumpf, New Brunswick 2001] 353–72), which covered such topics as how to measure the height of mountains, the sphericity of the earth, the oblong shape of the inhabited world, and the tides, all of interest to Eratosthenes. The history of Ephoros of Kyme, probably published in the 330s BC (G. L. Barber, *The Historian Ephorus* [second edition, Chicago 1993] 12–13) was notable for its section on world geography, the earliest example of an historian needing to include this topic as a discrete part of an historical work. Although ironically Ephoros was writing merely a decade before the great proliferation of geographical knowledge due to Alexander the Great and others of that era, he carried geographical inquiry as far as possible for his day, with some of the first detailed material on the Keltic and northern parts of the earth (Barber [supra] 28–30; *FGrHist* #70, F131–4). Ephoros' four books on geography may have been a significant inspiration to Eratosthenes.

F2. *Summary*: **Poets entertain rather than teach, yet at the same time are generally geographically astute, especially Homer. His description of Greece includes excessive elaboration, but is useful nonetheless. But there are some things that are not essential for a poet to know, and too much detail can become burdensome.**

Commentary: Eratosthenes began his *Geographika* with a discussion of Homer, a technique followed by the major Hellenistic geographers (Hipparchos F1–3 [see Dicks, *Hipparchus* 38], Poseidonios [see Kidd, *Commentary* 766–7], and Strabo [1.1.2]). Yet Strabo's intense objection to Eratosthenes' view of Homer means that little of the fragment actually comes from Eratosthenes. His primary point seems to be that the goal of a poet is amusement or entertainment rather than to be didactic. This is a standard Greek view from at least the sixth century BC, when

Xenophanes of Kolophon began to reject the traditional poetical claim to wisdom. While still showing respect for Homer (Xenophanes F10, in Xenophanes of Colophon, *Fragments* [ed. J. H. Lesher, Toronto, reprint 2001]), Xenophanes was among the first to point out his inadequacies (see his F11–12, and 14–16). Eratosthenes' view reflects Aristotle (the phrase "character, emotion, and actions" is a direct quotation of the opening of the *Poetics*) and his student Aristoxenos of Taras, the major ancient writer on music. It is more nuanced than Xenophanes' polemic, and more semantically and even philosophically oriented (what is the difference between amusement and teaching?), attempting (unsuccessfully, according to Strabo) to avoid the trap of being criticized for libeling Homer. The entire issue of the reliability and relevance of Homer was a major concern of Hellenistic scholars, although this controversy reached its peak after Eratosthenes in the second century BC. Those who upheld the literal value of Homer resorted to emendation as necessary (see F8). Eratosthenes, taught by the Stoic Zenon of Kition (Strabo 1.2.2), was caught between the two views of Homer: the Stoic idea that he was the ultimate repository of knowledge, and the expanding horizon that demonstrated the limitations of Homeric geography (see F. W. Walbank, *Polybius* [Berkeley 1972] 125–7; also Walbank, *Commentary* vol. 3, pp. 577–8; James S. Romm, *The Edges of the Earth in Ancient Thought* [Princeton 1992] 185–7).

Eratosthenes felt that despite the lack of scholarly depth of Homer and the early poets, they were in their own way geographically astute. He had little sympathy for the descriptive formulas of Homeric toponyms, finding them excessive, but not useless. Strabo saw all this as contradictory thinking, failing to see, or ignoring, that Eratosthenes was making a distinction between the geographical and poetical value of the epithets: to say that Haliartos is grassy (*Iliad* 2.503) serves a poetic function but does not add to the geographer's understanding of the place. Eratosthenes, whose reputation as a poet and philologist surpassed his as a geographer, believed that it was not necessary for a poet to be skilled in geography, or other disciplines such as farming or military practices, and that such knowledge would actually over-burden the poem and make it less attractive. These views may not be so much comments against Homer but the didactic poetry that was becoming common in Eratosthenes' own day. The *Phainomena* of Aratos of Soloi, a work that appeared in Eratosthenes' youth, touched on quasi-geographical subjects and to some extent farming (under the rubric of weather phenomena, lines 752–1154), a topic Eratosthenes specifically said was unnecessary for poets. Even the *Aitia* of his compatriot and mentor

Kallimachos may have had too much historical detail for his taste. But the supreme example of a poem overloaded with geographical material was the *Argonautika* of Eratosthenes' predecessor as Librarian, Apollonios. Given that the replacement of the latter with the former probably represented a shift in academic policy at the Library, Eratosthenes may have felt that geography had no place in contemporary poetry (although his own *Hermes* included some geographical lore). Yet he retained a sympathy for the use of geographical data by Homer and other early poets.

F2 also introduces an important figure in the recension of Eratosthenes' geographical scholarship, Hipparchos of Nikaia, used regularly by Strabo as a source. Hipparchos' three-book critique was titled *Against the Geography of Eratosthenes*, published perhaps half a century after Eratosthenes' death. It was the first full response to Eratosthenes and his methodology, and was by no means favorable. Modern knowledge of Hipparchos' treatise relies almost totally on Strabo's quotations of it, and thus Eratosthenes, Hipparchos, and Strabo have become impossibly tangled (see Dicks, *Hipparchus*).

F3. *Summary*: Homer only discussed nearby places, essentially those within Greece. His account of the wanderings of Odysseus is not accurate geographically, and neither are the commentators on it.

Commentary: Eratosthenes felt that Homer only talked about places relatively close to the traditional Greek world, a point Strabo again disputed, but certainly true. The issue is developed at somewhat greater length in F8. Locales that seemed far away to Homer (e.g. Ogygia, Kalypso's island [*Odyssey* 7.244], described as ἀπόπροθεν, "far off") were in fact relatively nearby in the expanded world of Eratosthenes' day: in his time it was believed that Kalypso had lived on one of the islands off Crete or around Malta. Even that most distant of places, the Blessed Isles (Hesiod, *Works and Days* 166–73), was at first located in the Mediterranean. In Hellenistic times the geographical significance of the toponymic information of Homer had become a matter of scholarly interest. Kallisthenes of Olynthos, the nephew of Aristotle, wrote a treatise on the *Catalogue of Ships* (Strabo 12.3.5) that seems to have expanded the world of Homer. Yet Eratosthenes was probably most influenced by his own teacher, Zenon of Kition, who was greatly concerned about the validity of detail in the Homeric poems (Dio Chrysostom 53.4). Other of the unnamed commentators on Homer may have included Theagenes of Rhegion, of the latter sixth century BC, and Kleanthes of Assos, an older contemporary of Eratosthenes and fellow student of Zenon (Strabo, *Géographie* 1.1 [ed. Germaine Aujac and François Lasserre, Paris 2003] 186). The wanderings of Odysseus spread through a wider world as topographical knowledge expanded (A. T. Fear, "Odysseus and Spain," *Prometheus* 18 [1992] 19–26).

F4. *Summary*: **One should not critique poetry in a scholarly or historical manner.**

Commentary: Because Eratosthenes' ideas were expressed negatively by Strabo, it is difficult to determine what Eratosthenes actually said, but, as shown in F2, it seems that he believed that one should not seek scholarly accuracy, especially of an historical nature, in the Homeric poems, but value them as literature. Since Eratosthenes was a poet himself, one would expect that his assessment of poetry was a strongly held personal and professional conviction, but his poetry is so little known it is difficult to analyze it. He wrote an *Erigone*, recounting the tale of the Attic heroine who committed suicide upon learning of the death of her father Ikarios: the poem won the praise of the author of *On the Sublime* (33.5; see Geus, *Eratosthenes* 100–10). He also wrote a poem in honor of Ptolemaios III, his inaugural lecture upon becoming Librarian, preserved in a late-antique commentary on Archimedes: see Geus, *Eratosthenes* 133–8. His *Hermes* recounted the youth of the god. Yet strangely all these poems have the didactic overtones that Eratosthenes seems to have condemned: the *Erigone* had an astronomical component (the principals became constellations), the inaugural poem solved the problem of doubling the cube, and the *Hermes* described the terrestrial climate zones.

F5. *Summary*: **The wanderings of Odysseus cannot be assessed in a scholarly fashion.**

Commentary: Eratosthenes stressed that the wanderings of Odysseus were to be taken metaphorically. The allusion to Homer is not exact: it was Aiolos himself who sewed up the bag of winds (*Odyssey* 10.19–20), something Eratosthenes would have well known, so the subtle point is made not to seek exact detail in the attractive tales of Homer and the early poets. The fragment introduces another important figure in the recension of Eratosthenes' geographical scholarship, Polybios, who was born about the time Eratosthenes died. He wrote a history of Rome's spreading control over the Greek world in the second century BC, events in which he played a part. Geography was an important part of this work, and he also explored remote places in West Africa and northwest Europe (Roller, *Pillars* 99–104). Like Ephoros, he included a section on world geography in his history, although this portion is only preserved in fragments, many of which are in the *Geography* of Strabo. He was not particularly well disposed toward Eratosthenes, condemning his lack of knowledge of the west (an area in which Polybios had special expertise) and affected by his own personal bias against Pytheas of Massalia, who was Eratosthenes' primary source for the west of Europe (F14). As was the case with Hipparchos, Strabo tangled Polybios' views in with those of Eratosthenes, so the usual problems of source transmission are apparent.

F6. *Summary*: **Homer, like all poets, tells falsehoods and elaborates excessively, not necessarily drawing on personal experience. This is especially apparent in the wanderings of Odysseus, where many of the places mentioned are fabricated, or at least elaborations. Any attempt to substantiate the topography of the wanderings creates contradictions. For example, the region around the Gulf of Kyme, which is associated with the Seirenes, is not accurately represented. Hesiod, knowing that the wanderings were in Sicily and Italy, added to the Homeric account, using toponyms unknown to Homer, and provided a greater degree of localization.**

Commentary: According to Eratosthenes, the wanderings of Odysseus (presumably considered as a general example of Homeric topographical concerns) could be examined in two different ways. One was to make a literal interpretation, but with poetic elaboration, and the other was to see them existing in a world of constructed fantasy. Eratosthenes felt that all poets elaborated even to the point of falsehood, but this was really irrelevant to appreciation of the poetry. Yet at the same time he understood that mythology could relate to a real topographical world: three examples are provided (Ilion, Ida, and Pelion), but whether these are from Eratosthenes or Strabo is not clear. But there was much that was invented about myth, and Eratosthenes believed that this, for the most part, was the case with the wanderings, because any attempt at topographical deconstruction would become mired in contradiction. Strabo provided a lengthy discussion of topographical issues around the Bay of Naples (a region that he knew well), describing in detail the long peninsula that marks its southern edge, where Sorrento is today, and indulging in some semantic quibbles about toponyms. Much of this may not originate with Eratosthenes, for there is no evidence that he knew the region nearly as well as Strabo. The ultimate issue of the argument is the location of the home of the Sirens, topographically undefined by Homer (*Odyssey* 12.39–54). In Eratosthenes' day there were several places associated with them: Pelorias (the eastern tip of Sicily), the rocks called the Seirenoussai, and the north side of the Sorrento peninsula. Eratosthenes favored the latter, although the Seirenoussai, whose ending shows that it is a Phokaian toponym (Jean-Paul Morel, "Les Phocéens en Occident: Certitudes et hypothèses," *PP* 21 [1966] 385–7), had the advantage of antiquity, yet their location outside the Bay of Naples was a problem. Pelorias was assumed to be close to Skylla and Charybdis, Odysseus' next peril. The fragment does not give the origin of the three suggestions, but Eratosthenes used their divergence to indicate that Homeric topography was unreliable in detail, and that such detail was unnecessary to appreciation of the poem. Strabo argued exactly the opposite, that since all three suggested locations were in the same region, this proved the validity of Homer. On the Sirens, and Odysseus in Italy generally, see E. D. Phillips, "Odysseus in Italy," *JHS* 73 (1953) 53–67.

Eratosthenes further believed that Hesiod built on Homer's topography and thus represented a wider world. In Eratosthenes' day there was still argument as to whether Homer or Hesiod was earlier (Aulus Gellius 3.11), although the balance was tilting toward the view held today, giving Homer precedence, and Eratosthenes, following Ephoros, agreed. Eratosthenes provided several Italian toponyms to demonstrate that Hesiod functioned in a broader environment, such as Aitna (*Theogony* 860), Ortygia (not the Ortygia of *Odyssey* 15.404, which seems to be in the east, but, as Strabo made clear, the island off Syracuse), and the Etruscans (*Theogony* 1016). The *Theogony* would have been a work of special interest to Eratosthenes because of its unusually broad geographical range: the catalogue of rivers at 337–45 is the earliest mention of the Nile, Istros, Phasis, and others. Unfortunately, if Eratosthenes devoted any detail to Hesiodic geography, Strabo did not preserve it. Implicit in these comments about Hesiod and Homer was Eratosthenes' realization that geographical fantasies are always placed just beyond human knowledge (see F7), something that has been true from ancient to modern times. On this passage see Strabo, *Géographie* 1.1 (ed. Germanine Aujac and François Lasserre, Paris 2003) 188–90.

F7. *Summary*: **Homer wished to locate the wanderings of Odysseus in the west, but did not know enough about the region to**

do it accurately, elaborating what little data he had and realizing that it is safe to be inaccurate about distant locales, the normal places for marvels.

Commentary: Strabo provided a list of places around the Bay of Naples associated with the wanderings of Odysseus, using them as proof that they had occurred in this region, especially noting that Baios and Misenos were companions of Odysseus who were reflected in well-known toponyms. The text is somewhat uncertain: see Strabo, *Géographie* 1.1 (ed. Germanine Aujac and François Lasserre, Paris 2003) 106, 191. What Strabo conveniently ignored is that neither of the companions was mentioned by Homer: they first appear, as do other of the names mentioned in this passage, in the *Alexandra* (lines 687–737) attributed to Lykophron. Although there were probably earlier sources, they are undoubtedly post-Homeric, and Strabo relied on a version probably even later than Eratosthenes. Regardless of the controversies over the date of the *Alexandra* (ranging from the late fourth century to the early second century BC, and even with Augustan interpolations: see Stephanie West, "Notes on the Text of Lycophron," *CQ* 33 [1983] 114–35), the fact remains that Strabo was critiquing Homer from a Roman perspective meaningless to Eratosthenes. Eratosthenes believed that Homer lacked sufficient knowledge about Italy and Sicily to allow more topographical accuracy, and (as in F6) it was wiser to use faraway places for miraculous events. Yet one is left wondering what the force of "this" (τοῦτο) is at the beginning of Strabo's Section 19. Did Eratosthenes also realize that in his time there was already close examination of Homeric tales occurring in Italy? There is no evidence that he was acquainted with emergent Latin literature (or Rome much at all, except as a topographical locus: see F60, 65), but he was a contemporary of Livius Andronicus, Naevius, and, to some extent, Ennius, all of whom were interested in applying a Roman perspective to these very issues. One might suggest that some of Eratosthenes' feelings about Homeric topography reflected the South Italian revisionism of his own era.

F8. *Summary*: **Homer tended to be ignorant about places outside Greece as well as long journeys by land or sea: there is more detail about places in Greece than, for example, the region of the Pontos or interior Africa. His ignorance of the Pontos was because it was not yet navigable. Although he does have certain ethnyms from this region, he does not mention the Skythians, and in general his knowledge is vague. His ignorance of Africa**

is extensive, unless one accepts Zenon's emendation of the *Odyssey*. Even more recent authors have been ignorant of distant places: these include Hesiod (who, however, did mention the Skythians), Aischylos, and many others. Inexcusably even Kallimachos has geographical inaccuracies.

Commentary: Apollodoros of Athens wrote a number of works in the second century BC that relied heavily on Eratosthenes. Of particular importance was his commentary on the Homeric *Catalogue of Ships*, which applied Eratosthenes' topographical researches to the vast number of toponyms in the *Catalogue*. The work only survives in fragments (*FGrHist* #244, F154–207), many from Strabo, who used Apollodoros extensively as a cross-check on Eratosthenes, often with relatively little critical comment. F8 is probably totally Strabo's summary of Apollodoros (his F157a). As in F6–7, there is the assertion that Eratosthenes believed Homer was reasonably reliable about the Greek world but less so regarding distant places. Eratosthenes provided a catalogue of places not mentioned by Homer, starting with the numerous rivers that flow into the Black Sea. It is perhaps not surprising that Homer would have been unaware of these since the Black Sea itself is not mentioned in extant literature before Herodotos (3.93 and elsewhere), and although there was Greek settlement along its coasts perhaps as early as the latter eighth century BC (beginning at the Milesian outposts of Sinope and Trapezos: see John Boardman, *The Greeks Overseas: Their Early Colonies and Trade* [fourth edition, London 1999] 239–43), detailed knowledge of the region is unquestionably post-Homeric. Even the early Greek trade with Kolchis at the southeastern corner, implied in the myth of the Argonauts, does not demonstrate any deep topographic understanding. The earliest references in Greek literature are to the "Kolchians," a vague ethnym associated with the Argonauts and lacking any precise localization. Eumelos of Corinth, a contemporary of Hesiod, may have been the first to cite the ethnym, but the source for this is late (Pausanias 2.3.10). Epimenides of Crete, slightly later than Eumelos, also wrote on the Argonauts (Diogenes Laertios 1.111). The first extant citations are not until the fifth century BC, beginning probably with Simonides (his F545), and then Aischylos, *Prometheus* 415 (the earliest use of the toponym as opposed to the ethnym) and Pindar, *Pythian* 4.11, 212. On early Greek knowledge of Kolchis, see David Braund, *Georgia in Antiquity* (Oxford 1994) 14–16. Hesiod was aware of the Phasis, one of the rivers flowing into the Black Sea (*Catalogue* 45), but again the isolated nature of this toponym does not imply much, and knowledge of the Black Sea remained vague until the seventh century BC at the earliest. This late opening of the Black Sea was eventually explained as due to its inhospitability to sailors, taking its original indigenous name, "aesaena" ("dark" or "somber," perhaps a Persian word) and transferring it into Greek "axeinos," "unfriendly," but later, after Greek penetration, it became "euxeinos," "friendly" (Stephanie West, "'The Most Marvellous of All Seas': The Greek Encounter with the Euxine," *G&R* 50 [2003] 151–67). The Skythians, the most remote of northern peoples, are perhaps first mentioned by Hesiod (Clement, *Stromateis* 1.16.75) but the reference is a late and dubious ethnym. Strabo was quick to point out that although Homer did not mention the Skythians themselves, he used ethnyms such as Hippemolgoi and Galaktophagoi (*Iliad* 13.5–6), peoples who were eventually associated with them, but Strabo ignored the fact that the geographical context of this passage was not the Black Sea but just across the Hellespont from Troy.

The next region in Eratosthenes' catalogue of places not mentioned by Homer is Egypt, specifically the Nile (and its peculiar characteristics). The river is first named by Hesiod (*Theogony* 338) and its flooding first discussed by Thales of Miletos (Aetios 4.1.1). It is perhaps unfair for Eratosthenes to complain about Homer's ignorance of the Nile floods, but it demonstrates how deeply the issue had become a part of Greek intellectual culture by Hellenistic times. Not cited in this passage, but another example of Homer's limited geographical vision, is Strabo's bald statement "Homer did not know about India" (1.2.32), a comment that may also have originated from Eratosthenes (see Klaus Karttunen, *India in Early Greek Literature* [Helsinki 1989] 107–8).

Because of Homer's seeming ignorance of the extremities of the world, there were many attempts to adjust his text, and an example is provided from Zenon of Kition, one of Eratosthenes' teachers. He took a line in the *Odyssey* (4.84) where Menelaos is describing his wanderings in the eastern Mediterranean, reaching the land of the Eremboi, an ethnym mentioned nowhere else by Homer. Zenon changed it to the more familiar Arabes, in part, one expects, because the Eremboi were obscure even in the early Hellenistic period. Little is known about them: the most detailed discussion, in fact, is another passage by Strabo (1.2.34–5), relying on Poseidonios (F280 Kidd), where there is a long and complex explanation suggesting that Eremboi is in fact an earlier variant of Arabes (thus upholding Zenon's claim), and that they lived somewhere in the Red Sea region. There seems no basis for this claim other than the similarity of the names, and the argument becomes quite circular and even convoluted. It is impossible to determine what the ethnym Eremboi meant to Homer and whether Zenon, Poseidonios, and Strabo were making any arguments other than linguistic ones: see further, Kidd, *Commentary* 954–6. The role of Eratosthenes in all this is even less clear, although Strabo's implication that most of the material in F8 is from Eratosthenes (through Apollodoros) would suggest that Eratosthenes had the same ideas.

Next Strabo listed a number of writers later than Homer who wrote about anatomically impossible people or fantasy worlds, again giving as his source Eratosthenes through Apollodoros. Part of this section may originally have come from the end of Eratosthenes' Book 1, where geographical fabrications were discussed (F19–24). The first half of the list cites several people whose names reflect their physical peculiarities, allegedly recorded by Hesiod, Alkman, and Aischylos, yet not appearing in the extant works of these authors, although one, the Pygmies, are in the *Iliad* (3.6), and Aischylos wrote about the "one-eyed Arimaspians" (*Prometheus* 804–5), not quite the same as the ethnym Monommatoi used by Strabo. It is probable that the list is from a catalogue of marvels. Following are nine topographical anomalies. Two are unlocated toponyms (Rhipaia, the "tossing mountain' of the far north [for which, see Jacques Desautels, "Les Monts Riphées et les Hyperboréens dans le traité hippocratique des airs, des eaux et des lieux," *RÉG* 84 (1971–2) 289–96], and Ogyion, cited nowhere else), and two are locales of mythical peoples (the homes of the Gorgons and the Hesperides). The remaining five are fantasy constructed places (or locations believed to be such). First is Meropia, a continent larger than the three known ones combined, described in the fourth century BC by Theopompos of Chios (*FGrHist* #115, F75c), an early example of the developing view that there were other continents beyond the known world. Hekataios of Abdera, also of the fourth century BC, wrote *On the Hyperboreans*, a fantasy treatment of the far north (*FGrHist* #264, F7–14), which included a city of the Kimmerians, a good example of how fantasy and real geography can become entangled, since the Kimmerians were an actual ethnic group, known to the Bible (Genesis 10.2 as Gomer) and Homer (*Odyssey* 11.14), whose movements affected Anatolia in the seventh century BC (Herodotos 1.16). Another fantasy world, Panchaia, was the creation of Euhemeros of Messene (*FGrHist* #63, F1–11), who wrote in the era of the Successors, and, as might be expected at that time, situated Panchaia beyond Arabia. The list continues with an otherwise-unknown reference from Aristotle to stones formed from sand but melted by rain, perhaps vague knowledge of the salt-block architecture of the Persian Gulf, first described by Androsthenes of Thasos (*FGrHist* #711, F2) in the 320s BC, an author well known to Eratosthenes (F94): the reference to Aristotle may be an error somewhere in the transmission down to Strabo. The final item is a Libyan city that cannot be found a second time, a stock element in folklore from ancient to modern times (such as in *Lost Horizon*). Significantly none of the authors cited is later than Eratosthenes, showing that he is the immediate source of the lists, although from the appropriate place near the end of Book 1 rather than in his discussion of Homer, but suggesting that while geographically astute he was no different from his successors in mixing reality and fantasy. Unexpectedly, however, the section closes with a libel of Kallimachos, a regular feature of the *Geographika* (see F2, 9, 12), who is described both as having only a pretense as a scholar, and in error about Homeric topography, another hint of the ongoing academic controversy between him and Eratosthenes. The remainder of the fragment is largely Strabo's refutation of Eratosthenes' attitude toward Homer, even using Eratosthenes' quotation of an otherwise-unknown line from Hesiod about "mare-milking Skythians" to prove that because Homer mentioned the Mare Milking People he also knew about the Skythians. However much Strabo turned the quotation to his own purposes, it shows that there was much more about Hesiod in the *Geographika* than has been preserved.

F9. *Summary*: **Kallimachos misunderstood Homeric topography.**

Commentary: Strabo used Apollodoros (for whom, see F8) to reinforce Eratosthenes' objection to Kallimachos' quality of scholarship. Since the alleged topographical error is the same as in F8, it is probably the same quotation from Eratosthenes, this time through Apollodoros. The error seems trivial, perhaps typical of major academic controversies. Kallimachos placed Kalypso's island at Gaudos (probably, but not certainly, the Maltese island of Gozo) and assumed that Scheria was Korkyra (Corfu). Yet the views of Eratosthenes and Apollodoros have become tangled, since it is Apollodoros, not Eratosthenes, who is referred to in the last clause of the passage ("he [Apollodoros] says"). The issue reflects the Hellenistic controversy about the wanderings of Odysseus: even if they were accepted as true (something Eratosthenes had reservations about), it was asked whether they were within or outside the Mediterranean (ἐξωκεανισμός, a word that became a virtual polemic in the controversy). The argument actually seems to have reached its greatest intensity after Eratosthenes, in the second century BC, with Krates of Mallos (outside) and Polybios (inside) the major proponents (see Walbank, *Commentary* vol. 3, pp. 577–8). Even though Eratosthenes is cited in the fragment as in disagreement with Kallimachos, the passage as a whole reflects the second-century world of Apollodoros.

F10. *Summary*: **Homer, unlike Hesiod, did not know the name "Nile" or how many mouths the river has. Homer was also in error about the island of Pharos and the isthmus between the Mediterranean and the Red Sea.**

Commentary: In the *Odyssey*, the Nile is called simply the "Aigyptos River" (14.258, 17.427): the name "Nile" first appears in the catalogue of rivers in Hesiod's *Theogony* (338). Arguing that Homer did not know that there were several mouths to the Nile is picky (in fact there is no mention of mouths at all) but shows how incomprehensible it was to Eratosthenes that someone was ignorant of this obvious fact. Like F8, the passage hints at a greater analysis of Hesiod than was preserved by Strabo. Eratosthenes also criticized Homer's location of the island of Pharos: his single mention of the island (*Odyssey* 4.354–7) describes it as a day's travel from the coast, but with a good harbor where Menelaos spent 20 days waiting for a favorable wind. This in no way conforms to the place just off Alexandria that Eratosthenes knew well and which carried the name in later times. Strabo's suggestion (1.2.23) that Menelaos was merely exaggerating his difficulties has a certain merit. Eratosthenes' final point concerns the isthmus between the Mediterranean and the Red Sea—which Homer did not mention—and his error in dividing the Aithiopians into two groups. The two issues are not particularly connected (their juxtaposition may be due to Strabo's editing), and Homer's failure to mention the isthmus, and indeed the Red Sea, was pure ignorance, or because of irrelevance to his needs. The splitting of the Aithiopians (which Strabo believed was implicit mention of the Red Sea) is in fact the earliest example of the tendency to see the Aithiopians as falling in two groups, those on the upper Nile south of Egypt, and those everywhere west, the so-called Western Aithiopians. As knowledge of sub-Saharan and Atlantic Africa increased, beginning in the latter sixth century BC with explorers such as the Carthaginian Hanno and the traders who came to Kyrene described by Herodotos (2.32), the term Aithiopian was used for everyone in Africa other than the Egyptians and those living on the Mediterranean coast.

F11. *Summary*: **Homer's reference to Boreas and Zephyros as Thracian winds is not correct, since the Zephyros comes from as far west as Iberia.**

Commentary: At the beginning of Book 9 of the *Iliad* the Achaians are stricken with panic, which Homer described as like the turbulence of the sea affected by winds from Thrace. Eratosthenes took issue with the topography of the simile, pointing out that the north and west winds do not blow from Thrace, but from farther away. Although there is only Strabo's polemical word for what Eratosthenes actually said, the objection seems trivial and ignores the highly localized nature of Homer's comment. It may be that Eratosthenes used Homer's simile about the winds around the Gulf of Melas (the northeastern corner of the Aegean) to make a more general discussion about winds and their origins, commenting that the winds that affected the Aegean came from as far away as Iberia. Wind theory was still relatively new in Eratosthenes' day— Aristotle's *Meteorologika* is the first detailed study of the topic—and he may simply have been applying contemporary theory to a Homeric issue. A comment by the late Roman author M. Cetius Faventinus, that Eratosthenes was able to learn the directions of the winds from calculating the circumference of the earth (*de diversis fabricis architectonicae* 2), is a garbling of Vitruvius 1.6.9, and not any proof of research on winds on the part of Eratosthenes: see Hugh Plommer, *Vitruvius and Later Roman Building Manuals* (Cambridge 1973) 88. Aristotle believed that winds were caused by solar heating of the earth and its atmosphere, and knew that they could travel some distance (*Meteorologika* 2.4). On this fragment see Kidd, *Commentary* 515–22; and A. Thalamas, *La gèographie d'Ératosthène* (Versailles 1921) 180–5, with some speculation about Eratosthenes' theory of winds.

F12. *Summary*: **The first two successors of Homer were Anaximandros, who was the first to make a geographical map, and Hekataios, although there is some uncertainty about the authorship of his geographical work.**

Commentary: Having completed his discussion of Homeric geography, Eratosthenes turned to his successors, Anaximandros and Hekataios, both from Miletos. Yet there is no extant discussion of either, probably due to Strabo's editing, and one can only surmise what Eratosthenes might have said about them. Anaximandros was active in the first half of the sixth century BC (Diogenes Laertios 2.1–2) and was primarily remembered for his cosmology. Relevant to Eratosthenes would have been his reputation as the first to theorize about the shape of the earth, believing that it was in the form of a column drum (Duane W. Roller, "Columns in Stone: Anaximandros' Conception of the World," *AntCl* 58 [1989] 185–9), a view that would have seemed exceedingly primitive and outdated in Eratosthenes' day. More significant was the belief that he was the inventor of mapmaking (supra, pp. 3–4). Both these accomplishments would have earned Anaximandros an important place in the history of geography. Hekataios' extensive *Circuit of the Earth*, written around 500 BC and surviving in numerous fragments (*FGrHist* #1) was the first of its type. Yet in Eratosthenes' day there was a dispute about whether the extant text was a legitimate work of Hekataios, as Kallimachos had doubted the authenticity of the Asian portions, suggesting that it was the work of a certain Nesiotes, otherwise unknown (Athenaios 2.70a–b [the name Nesiotes may even be a textual corruption]; see further Lionel Pearson, *Early Ionian Historians* [Oxford 1939] 31–4). Whether this controversy referred generally to Hekataios' writings or merely the copy in the Alexandria library is not clear, but since Eratosthenes rejected Kallimachos' assertion, it is yet another example of the academic controversy between the two (see F8–9).

F13. *Summary*: **Damastes recorded the tale about the questionable journey of Diotimos, who sailed from Kilikia to Sousa, a story that has topographical difficulties but which may reflect the lack of known sea routes. Yet many sources are nonsense, such as Antiphanes of Berga, Euhemeros of Messene, and others. One should be careful about the sources one believes, especially those concerning remote places. Even today much is still unknown about coastlines. The easternmost point of the Mediterranean is at the Issic Gulf. The northern and eastern Adriatic are not well known. There is some information about places beyond the Pillars of Herakles, including the island of Kerne. Early seamanship was coastal, and existed for reasons of piracy and commerce. Jason abandoned his ships in Kolchis and went overland as far as Armenia and Media, but in early times no one sailed on the Black Sea or the eastern and southern Mediterranean coasts.**

Commentary: Damastes of Sigeion was active in the latter fifth century BC. Only a few fragments survive of his geographical work (*FGrHist* #5), although they indicate a wide range. Eratosthenes quoted him at least twice (see also F130), here with the peculiar tale of the Athenian Diotimos (probably the leader of a naval force to Korkyra just before the Peloponnesian War [Thoukydides 1.45]), who led an embassy to Sousa by sailing up the Kydnos River (at Tarsos) and using various streams to reach his goal 40 days later. Although the tale had the validation of a personal report by Diotimos to Damastes, it is geographically impossible, something Eratosthenes realized. The itinerary may simply reflect Diotimos' route, not that it was done all the way by a single ship or that all the necessary rivers connected. It is improbable that he would have taken a sea route (around the Arabian Peninsula) because such was unknown in the fifth century BC. Strabo also criticized Eratosthenes for reporting Damastes' belief that the Arabian Gulf (the Red Sea) was a lake, but again this is perfectly reasonable for the fifth century, when it had hardly been explored by Greeks. It was not until the Augustan period that the coastal routes around Arabia came to be fully known (M. Cary and E. H. Warmington, *The Ancient Explorers* [revised edition, Baltimore 1963] 73–87).

Eratosthenes' use of Damastes led Strabo to complain about two other allegedly dubious sources. One is Antiphanes of Berga, a probable contemporary of Eratosthenes who wrote about the frozen north (Plutarch, *How One Might Become Aware of His Progress in Virtue* 7), known as a fanciful author. Strabo mentioned him dismissively several times as merely the "Bergaian" (Strabo 2.3.5, 2.4.2). The other is Euhemeros of Messene, active around 300 BC, who wrote a fantasy about the lands beyond Arabia (*FGrHist* #63, F1–11). Yet Strabo failed to realize that most fantasy authors contain within them actual topographical details, although it is not known how sensitive Eratosthenes was to this.

In a further attempt to discredit Eratosthenes, Strabo listed a number of perceived geographical errors. In most cases these represent the changes in knowledge from the third century BC to the Augustan period, such as the scant treatment of the upper Adriatic, on the fringes of the world in Eratosthenes' day but central to the Roman world of Strabo's time. Strabo also included what he saw as an outright topographical mistake, arguing that Eratosthenes had said that the Gulf of Issos was the easternmost part of "our sea," when it was actually the coast of the Black Sea near Dioskourias. There is no doubt that Dioskourias is over 350 km. east of the longitude of the Gulf of Issos, although even it is not the easternmost part of the Black Sea, which is south of the mouth of the Rioni, about 70 km. farther east. Eratosthenes knew the relative positions of Dioskourias and the Gulf of Issos, so Strabo's point is probably semantic: does one consider the Black Sea part of "our sea"? Strabo also criticized Eratosthenes for his data from beyond the Pillars of Herakles, as he mentioned a number of places that, to Strabo at least, were unknown. This actually reflects a regression of geographical knowledge in late Hellenistic times. Kerne, the only toponym mentioned, was a Carthaginian trading post in West Africa, founded by the explorer Hanno around 500 BC. A summary of Hanno's report was promptly translated into Greek, which was perhaps available to Eratosthenes (E. H. Bunbury, *A History of Ancient Geography* [second edition, London 1883] vol. 1, pp. 330–1). In the mid-fourth century BC a Greek traveler reached Kerne and found it in decline (Pseudo-Skylax 112). Kerne continued to become less viable: when Polybios visited it just after the fall of Carthage (34.15.9) he may have found it abandoned. It had been essentially forgotten by the early Augustan period. Other Carthaginian outposts in West Africa, presumably named by Eratosthenes, also faded away (at some uncertain date, perhaps as early as the fourth century BC, 300 had been deserted [Strabo 17.3.3]).

The fragment closes with a discussion of early seamanship. Eratosthenes argued that in ancient times there was little sailing, and that it was localized, for commercial and piratical reasons. This meant that the long Homeric journeys had to be rejected, especially those of Odysseus. Even Jason and the Argonauts were presented in a diminished capacity: they abandoned their ships and went overland part of the way, a variant on the traditional myth that may have gained popularity at the time of Alexander in order to make a comparison between hero and king (Strabo 11.4.8, 11.5.5). As usual, Strabo had little sympathy for Eratosthenes' view.

F14. *Summary*: **Pytheas traveled throughout Brettanike and went to Thoule and other places that had peculiar phenomena, although how much he is to be believed about the more remote areas is debatable. On the other hand, Euhemeros, who went only to Panchaia, is a fanciful source.**

Commentary: The immediate source of this fragment is Polybios, but for the most part it is about Pytheas of Massalia, one of the most difficult sources for the modern reader to understand. In the 320s BC Pytheas made an astounding journey from his hometown to northwest France and then across to the British Isles. He spent some time there, and then continued north to the Faeroes and the place that he called Thoule, almost certainly Iceland, where he recorded the unusual characteristics of the region, especially its mixture of glacial and volcanic phenomena, which made it appear that the world was not fully formed. He then returned to the European coast somewhere in the vicinity of Bergen in Norway and penetrated the Baltic to its eastern end. He may even have gone overland to the Black Sea. Throughout the journey he was an avid scientific observer, recording data about the tides, latitudes, Arctic phenomena, and the location of the celestial pole. When he returned to Massalia he published his travels in the treatise *On the Ocean* (ed. Roseman [Chicago 1994]; Roller, *Pillars* 57–91). Needless to say such a significant, far-ranging, and amazing journey was more disbelieved than understood. Eratosthenes, although ambivalent about Pytheas, used his data for the far west and north and incorporated Thoule into his system of parallels (F34–5, 37), but belief in Pytheas' veracity became, over time, less accepted. This attitude reached its peak with Polybios, as can be seen in F14, who seems to have had a particular distaste for the Massalian, perhaps because he saw him as a threat to his own reputation as an explorer. Unfortunately Strabo relied almost totally on Polybios' view of Pytheas (his treatise may no longer have been extant), which means that not only did Strabo validate the negativism toward Pytheas, but eliminated from his synthesis of Eratosthenes much of the latter's data obtained from Pytheas, which was most of what Eratosthenes reported about the north and west of Europe. Instead, Strabo reported that Eratosthenes was ignorant of those regions (F132). Anything that Eratosthenes obtained from Pytheas has become so badly corrupted at the hands of Polybios and Strabo that it is almost impossible to retrieve. Indeed Strabo, following Polybios, suggested that the fantasy geography of Euhemeros (see F13) was more reliable than Pytheas, on the strange argument that Euhemeros went to only one place but Pytheas to many. The entire passage is more tangled than usual for Strabo, and the source of particular topics can be impossible to determine with certainty.

Also mixed into the account are comments about Dikaiarchos of Messana (see F1), who himself rejected Pytheas, according to Eratosthenes. In a moment of particular astuteness, Strabo pointed out that neither Eratosthenes nor Dikaiarchos had been in northwest Europe so it was understandable that their data might be in error, criticizing Polybios for insisting that they were less than authoritative.

F15. *Summary*: **After the time of Alexander, great advances were made in the knowledge of the inhabited world. The world itself is spherical and has been subject to numerous natural processes that change its shape, although nothing affects the overall**

spherical form. As an example, there are marine phenomena well inland, such as at the Temple of Ammon, which once may have been on the sea. Such phenomena were reported by Straton and Xanthos; Straton also discussed changes in the sea levels and the flow from one sea to another. There are many other examples of sea levels that were different in earlier times from what they are today.

Commentary: Eratosthenes recognized that in the century since Alexander much that was new had been learned about the inhabited world. If he specified details, they were not preserved by Strabo, but there are a number of hints in Book 3 (F67–81) of the kind of information to which he refers, especially about the world east of Mesopotamia. Most of F15, however, discusses the shape of the world and its geological history, first establishing that Eratosthenes believed in a spherical earth, the Pythagorean concept that had received legitimacy through Plato, although rather vaguely expressed by him (*Phaidon* 58). The idea was still new enough that Eratosthenes felt it was necessary to make his belief in it explicitly clear. At the same time he asserted that the earth was not a perfect sphere (thus distancing himself from the Pythagorean view [Aristotle, *On the Heavens* 2.13]), but noting that despite the surface irregularities and degradation due to natural phenomena the basic spherical shape was valid. Eratosthenes' one documented field trip, to the site of the Achaian earthquake of 373 BC (F139), shows his particular interest in the changes to the surface of the earth. But especially intriguing to him was the presence of ocean phenomena far from the seacoast. Greeks had long been aware of the evidence that the surface of the earth had been radically different in the past. Herodotos (2.12) had noted seashells in the uplands of Egypt, which led him to thoughts about erosion and deposition, as well as the immense age of the earth. Eratosthenes focused his interest on the oasis of Ammon, west of Egypt, where maritime phenomena were common despite its distance (about 300 km.) from the sea. The famous oasis, occupied since prehistoric times, had become prominent in the early sixth century BC. By the fifth century it was thoroughly in the mainstream of Greek culture: Kimon of Athens consulted the oracle just before he died (Plutarch, *Kimon* 18), which means it was well respected (at least among the Athenians) by the middle of the century. When Alexander the Great visited the site in 331 BC (Arrian, *Anabasis* 3.3–4; the best ancient description is by Diodoros, 17.49–51), it was vividly thrust into the Greek consciousness. Early travelers would have seen the extensive marine phenomena (for an account from the early twentieth century of the several-day journey from the Nile, with seashells encountered all along the route, see C.V.B. Stanley, "The Oasis of Siwa," *Journal of the Royal African Society* 11 [1912] 290–324; for an engaging discussion of the oasis from ancient to modern times, see Ahmed Fakhry, *Siwa Oasis* ([Cairo 1973]). The site lies in a depression below sea level, with the ground rising to the north, toward the Mediterranean, reaching the highest elevation (less than 300 m.) near the coast, before dropping sharply through the Katabathmoi ("The Descents") to the sea. The list of maritime phenomena visible at Ammon has a touristic quality: although some are legitimate, it seems unlikely that there were remains of seagoing ships, and as it stands an inscription from Kyrenaian ambassadors proves nothing, as Kyrene had had a close relationship with the shrine since the founding of the city (Ammon appears on its coins from at least the latter sixth century BC: see Barclay V. Head, *Historia Numorum* [Oxford 1911] 865–7; also C. J. Classen, "The Libyan God Ammon in Greece before 331 BC," *Historia* 8 [1959] 349–55). The use of "are shown" (δείκνυσθαι) implies a visit to a tourist trap (but see further, F16).

Eratosthenes then cited his two major sources for the formation of the earth. Xanthos, from Sardis, wrote a logographic history of Lydia in the fifth century BC, titled *Lydiaka* (*FGrHist* #765), which seems to have had an unusually strong focus on geology. There are discussions of the siltation of rivers in western Anatolia and volcanic phenomena (Xanthos, F13; see further Lionel Pearson, *Early Ionian Historians* [Oxford 1939] 109–38) in addition to the passage quoted by Eratosthenes (Xanthos, F12), which mentions a major contemporary drought (the King Artaxerxes must be the first of that name, 465–424 BC), although its location is not specified. Xanthos' interest in drought may have come from his possible teacher Empedokles of Akragas (whose biography Xanthos wrote [(Diogenes Laertios 8.63]), probably the first to discuss the phenomenon (Empedokles, F111), but in a manner more mystical than scientific. Xanthos also recorded visible seashells in a wide sweep from central Anatolia to the Kaukasos (if ἰδεῖν is to be taken literally, he was an exceedingly broad traveler), which led him to believe that the plains were once the sea. One can argue whether any of the territory specified qualified as plains (τὰ πεδία), but significantly everywhere is over 1000 m. above sea level.

The other source that Eratosthenes used at this point was Straton of Lampsakos, who was Theophrastos' successor as head of the Lyceum and died about the time Eratosthenes came to Athens (see H. B. Gottschalk, "Strato of Lampsacus," *DSB* 13 [1976] 91–5). He was committed to a physical universe that operated by natural processes rather than divine causation (see Plutarch, *Reply to Kolotes* 14; Cicero, *de natura deorum* 1.35), which led him to suggest that the earth had been subject to many changes. Eratosthenes' lengthy paraphrase of Straton's views concerns the levels of the seas, especially the idea that the exit from the Mediterranean at the Pillars of Herakles had not always existed. Strikingly, he seems to have been well informed about the depth of the seas, including the relative shallowness of the western Mediterranean and the great depth of the waters west of Italy, as well as the effect of river sediments, one of several ancient attempts to determine sea depth. Poseidonios (F221 Kidd = Strabo 1.3.9) recorded that the Sardinian Sea (that west of the island) was 1,000 *orgyiai* deep, which would be about 2,000 m., well short of its actual 3,000 m. How the depth was determined is unknown (Kidd, *Commentary* 794–5).

Straton was probably Eratosthenes' source for the belief that the depression at the oasis of Ammon had once been part of the sea. There was also a presumption that the sea levels around Egypt had been higher in earlier times. Straton built on ideas originally expressed by Aristotle (*Meteorologika* 2.1–2) and concluded that the original outlet of the Mediterranean had been across the isthmus to the Red Sea, where there remained a large amount of marine phenomena and the land was no more than 80 m. above sea level and frequently below it (see also F16). The many strong rivers emptying into the Black Sea caused it at some early date to burst into the Aegean, which forced the Mediterranean itself through the Pillars of Herakles rather than exiting into the Red Sea. There was even a mythological memory of this event, as the inhabitants of Samothrake recorded a great flood that had caused water from the Black Sea to inundate much of the island, resulting in architectural debris that fishermen were still recovering (Diodoros 5.47.4–5). Straton detected a flow in the Black Sea from its major river estuaries that created the strong current (3–5 knots) at the Hellespont, an explanation still sustained today (Benjamin W. Labaree, "How the Greeks Sailed into the Black Sea," *AJA* 61 [1957] 29–33). He also believed that the bed of the sea was uneven, just like the surface of the earth, because it had once been land. On the issues of physical geography in F15–16, see Germaine Aujac, "Ératosthène et la géographie physique," in *Sciences exactes et sciences appliquées à Alexandrie* (ed. Gilbert Argoud and Jean-Yves Guillaumin, Saint-Étienne 1998) 247–61. In recognition of Eratosthenes' research on the bed of the Mediterranean, features just south of Cyprus have been named the Eratosthenes Seamount and the Eratosthenes Abyssal Plain.

F16. *Summary*: **Despite what Archimedes asserted in his *On Floating Bodies*, the levels of the Mediterranean are not the same—even though it is a single sea—as engineers have shown at the Isthmos of Corinth. In narrow straits there are strong currents and changing tides, which cause differences in sea levels. In fact, each sea has a different level. The Mediterranean was once connected to the Red Sea, but later became connected instead to the External Ocean at the Pillars. All the seas flow together, although this does not mean that they are all at the same level. These changes in the Mediterranean probably caused the flooding at the Temple of Ammon, and even perhaps flooding from the Black Sea to the Adriatic via the valley of the Istros.**

Commentary: In his treatise *On Floating Bodies* Archimedes put forth the following proposition (Book 1, Prop. 2): "the surface of all fluids that remain immovable will have the surface of a sphere having the same center as the earth." Eratosthenes, more practical than mathematical, would not accept the conclusion of his friend and academic colleague (supra, p. 12), noting somewhat pedantically that the level of the Mediterranean varied, especially within its various gulfs and bays. Eratosthenes discussed in detail the levels and currents of the Mediterranean, arguing that engineers were better informed than mathematicians. A canal at the Isthmus of Corinth planned by Demetrios (almost certainly Poliorketes, who was based in Corinth in 302–1 BC) was not built, because his engineers told him that the project would create such a flow from the Gulf of Corinth to the Saronic Gulf that Aigina would be flooded. Although the engineers were correct about the flow (the current through the modern canal is 1–3 knots), this may have been their tactful way of suggesting that the project was ridiculous, as it would have required an eternity to construct a canal nearly six km. long and 90 m. deep (the French took 12 years, from 1881 to 1893, to build the present canal). Attempted canals through the Isthmos have had a long history, beginning with Periandros of Corinth around 600 BC (Diogenes Laertios 1.99; see the list in J. G. Frazer, *Pausanias's Description of Greece* [reprint, New York 1965] vol. 3, pp. 6–8), but Demetrios' was the most recent in Eratosthenes' time. His information may have been an oral report from someone involved in the project.

The comments on the proposed Corinth Canal led Eratosthenes to the matter of currents through straits, which almost immediately produced a further discussion, about tides. Tidal theory was a particularly complex issue in Greek intellectual culture, with understanding hampered to some extent by the limited tides of the Mediterranean. The early explorers who ventured into the Atlantic were baffled by them: there were reports of ships being stuck in the open sea (Sataspes, a Persian, between 479–465 BC: Herodotos 4.43) and rivers flowing inland from the ocean (Euthymenes of Massalia, about 500 BC: Aristeides 36.85–95). Pytheas of Massalia was the first to think carefully about the tides, as he encountered some of the largest in the world, around the British Isles, and made the important conclusion that tides were somehow connected to the moon (Aetios 3.17.3; Pliny, *Natural History* 2.217). He was probably Eratosthenes' source for tidal comments (see F153), but needless to say Strabo did not acknowledge this, and the reader is simply referred to sources later than Eratosthenes. It was not until the second century BC that an actual treatise was written on the tides, by Seleukos of Seleukeia (Duane W. Roller, "Seleukos of Seleukeia," *AntCl* 74 [2005] 111–18).

Eratosthenes also believed that the well-known phenomenon of the change of direction of the current through straits (the three most famous are cited, the Strait of Messina, the Euripos at Chalkis, and the Hellespont) was due to the differing levels of various portions of the Mediterranean (see Kidd, *Commentary* 783–4; Strabo, *Géographie* 1.1 [ed. Germaine Aujac and François Lasserre, Paris 2003] 209–10). This too was a persistently baffling problem that was never solved in antiquity. Aristotle was said to have pondered the matter of the Euripos in his last days at Chalkis, and his inability to solve the problem caused his early death (Prokopios, *History of the Wars* 8.6.20). If Eratosthenes further developed these questions, Strabo did not preserve it, as his main interest was in refuting both Eratosthenes' disagreement with Archimedes, and Eratosthenes' next point, the matter of a prehistoric connection between the Mediterranean and the Red Sea. This became caught up in a semantic argument regarding the meaning of Eratosthenes' word "connection" (τὸ συνάπτειν) and whether it really means "near" or "to touch." The quibbling is the result of a serious inconsistency in Eratosthenes' views, which Hipparchos (his F8) pointed out: since the External Ocean was continuous, why should the Mediterranean, even after receiving the outflow from the Black Sea, not continue to have an exit into the Red Sea? Strabo's analysis of both Eratosthenes and Hipparchos is less than clear, and his own ignorance about tides clouds his argument (although he was familiar with Seleukos' treatise, through Poseidonios [Strabo 3.5.9]).

As the text stands, the matter of the Kyrenaian dolphins seems irrelevant to the levels of the sea. Yet the text is somewhat uncertain at this point, and editors from Berger (his pp. 57–9) on have found the matter peculiar. Obviously Eratosthenes was trying to make a point, even with the touristic overtones of the collection of artifacts shown to visitors to Ammon (F15). It has reasonably been suggested that the dolphins were not part of an inscription but from the shipwrecks mentioned in F15, reading, instead of στυλιδίων ("small columns"), στυλίδον, which has some manuscript authority and means "a mast to carry a flag at the stern," a word used elsewhere by Eratosthenes (*Katasterismoi* 35; see also Plutarch, *Pompeius* 24.3; for the emendation and the evidence for it, see Dicks, *Hipparchus* 119). This would give more sense to the Kyrenaian dedication, but it still requires the unlikely fact that Ammon had been on the seashore some time after the founding of Kyrene around 630 BC, which Hipparchos (his F9) did not believe. Strabo (or more probably, Hipparchos) pointed out another inconsistency: if Eratosthenes believed both the Mediterranean–Red Sea isthmus and Ammon were covered, much of Egypt and indeed many other places would also have been underwater. For example, the engineers of Ptolemaios II had determined that the Red Sea was (in Pliny's units, from *Natural History* 6.166), three cubits (ca. 1.3 m.) above the level of the Nile, and thus further above the level of the Mediterranean. The Black Sea would have flowed into the Adriatic, as it was believed that the Istros (Danube) split into two with one branch flowing into the Adriatic. This strange characteristic of the Istros is first documented in the *Periplous* of Pseudo-Skylax, of the mid-fourth century BC (see also Aristotle, *Research on Animals* 7[8].13), and is probably due to a vague misunderstanding of the complex ancient riverine trade routes of central Europe from the Baltic to the Mediterranean or the Black Sea, something alluded to by Herodotos (4.48) but which received its greatest popularity with Apollonios' rendering of the Argonaut tale (*Argonautika* 4), roughly contemporary with Eratosthenes. The idea may also have resulted from the fact that tributaries of the Danube are remarkably close to the Adriatic (less than 20 km. in Croatia) and the duplication of the name Istros (the Danube) and Histria (the modern Istria peninsula), presumably variants of a regional ethnym (Hekataios [*FGrHist* #1, F91]).

F17. *Summary*: **Before the Mediterranean was connected to the External Sea, it joined the Red Sea, but the present isthmus in Egypt was created when the Mediterranean joined the External Ocean at the Pillars.**

Commentary: This fragment essentially repeats data from F16, with the additional fact that Eratosthenes believed that the changes in the

Mediterranean and the elimination of its connection to the Red Sea had happened after the Trojan War. He evidently based this on *Odyssey* 4.84 (the same passage is discussed in F8), where Menelaos tells Telemachos that he visited the Aithiopians; line 82 implies a sea journey. A certain Aristonikos, Strabo's contemporary, had discussed this in his *On the Wanderings of Menelaos* and had recorded a number of inventive explanations for the voyage, including that of Eratosthenes. Strabo was quick to point out that there was no other evidence for this idea.

F18. *Summary*: **The changes discussed in F17 caused the Dead Sea.**

Commentary: The context and text itself are confusingly corrupt. Since 16.2.1 Strabo's description has been moving south through Syria and Phoenicia, reaching Judaea at 16.2.34, and moving inland to Jericho at 16.2.41, with mention of Lake Sirbonis at 16.2.42. Sirbonis is the lake on the eastern coast of Egypt (see F15), relevant to the arguments about the connection between the Mediterranean and Red Sea, but mention of Moasada (Masada) and Sodom (16.2.44) shows that Strabo was describing the Dead Sea. Its common ancient name, Asphaltitis, is not in the extant text of Strabo and does not appear in existing Greek literature until Josephus (*Jewish War* 4.437–9), although both Strabo and Diodoros (19.98) used it as a descriptive rather than toponymic adjective. Nevertheless it is impossible to tell how early the term came into Greek use. Aristotle, the first Greek to mention the Dead Sea (*Meteorologika* 2.3), provided no name. Strabo's immediate source for at least part of the passage is Poseidonios, who may have used the name Asphaltitis (see his F219 Kidd), but the source for this fragment is so late (Priscianus) that the nomenclature is hardly guaranteed. It quite probable that Strabo or Poseidonios transferred Eratosthenes' discussion of the sea levels around Lake Sirbonis to the Dead Sea, which is mentioned nowhere else in the fragments of Eratosthenes. In addition, at the end of the fragment Strabo's text has the uninformative "καθάπερ τὴν θάλατταν." His text was emended by A. Corais in 1819 to read "Θετταλίαν," following Herodotos' description (7.129) of the Thessalian plain having been a lake until Poseidon created the Vale of Tempe as an outlet (a less mythological explanation is at Strabo 9.5.2). Since there is no other mention of Thessaly in the fragments of Eratosthenes, it is undetermined whether he discussed the region. See Nicola Biffi, *Il medio oriente di Strabone* (Bari 2002) 244–5.

F19. *Summary*: **There are those who discuss fabrications, both mythological and historical.**

Commentary: The final section of Book 1 of the *Geographika* was about fantasy geography. This was a genre that had developed in the fifth century BC, at first serving allegorical or moral purposes. When it became obvious that the known world centering on the Mediterranean was only a small part of the surface of the earth, it was possible to imagine what other continents might be like, and to use such places to create ideal worlds. The Pythagoreans may have been the first to suggest the existence of other continents (Diogenes Laertios 8.25–6), but it was Plato who not only popularized the idea but created the first significant allegory about another world, Atlantis (*Timaios* 24e–25d). Theopompos (*FGrHist* #115, F75c), Hekataios of Abdera (*FGrHist* #264, F7–14), Euhemeros of Messene (*FGrHist* #63, F1–11) and Antiphanes of Berga (see F14) were later proponents of the genre whose writings were accessible to Eratosthenes. Yet as geographical knowledge expanded in the latter fourth century BC, actual data from remote places could be incorporated into the allegories, which made separation of fact and fantasy difficult, a problem continuing into modern times. Although Eratosthenes' focus may have been more on myth than these fantasy allegories (see F21), Strabo took him to task for believing in authors that Strabo did not, most notably Pytheas (F14), whose report of bizarre northern phenomena was fertile inspiration for the fantasy writers.

F20. *Summary*: **Herodotos' statement about the Hyperboreans is absurd.**

Commentary: The Hyperboreans first appear in the Homeric *Hymn to Dionysos* (line 29; see also Hesiod, *Catalogue*, F150), with the earliest description of their ideal environment by Pindar (*Pythian* 10.27–44). Although repeated attempts were made to find them, or to connect them with newly discovered peoples, their generic name ("People beyond the North Wind") indicates that they are fictional rather than geographical (James S. Romm, *The Edges of the Earth in Ancient Thought* [Princeton 1992] 60–7). Herodotos did not actually write as he is quoted (4.36), whether by Strabo or Eratosthenes, saying something quite different: that if there were Hyperboreans there must be Hypernotians ("People beyond the South Wind"). In all likelihood he was speaking ironically, but he thus raised the question of what existed at the southern ends of the earth, something more difficult to comprehend because all the known world was in the northern hemisphere. Eratosthenes, perhaps taking Herodotos' comments too literally, used a logical argument to reject it, thereby seeming to accept the existence of both Hyperboreans and Hypernotians, with a hint of knowledge of the southern east coast of Africa. The Pythagoreans had been the first to suggest a southern land mass (Diogenes Laertios 8.26), but this was an argument of mathematical symmetry, not geography, and there is no evidence that either Greeks or Romans got very far south of the equator, leaving the southern land mass essentially out of ancient comprehension, although early modern explorers fixed on these ideas and long looked for the *terra australis incognita*. The word "antarctic" only appeared in the first century BC (Geminos 5.16, 28–9), merely referring to the antarctic circle, and the continent of Antarctica, although long supposed, eluded explorers until the early nineteenth century.

F21. *Summary*: **The stories about the wanderings of Herakles and Dionysos are legendary.**

Commentary: It is probable that much of Eratosthenes' discussion of geographical fabrications related to mythology, rather than fantasy material in specific authors. The first travelers were the gods Herakles and Dionysos. Dionysos went from his Lydian home through Baktria, Media, Arabia, Asia, and eventually to Thebes in Greece, bringing the grapevine (Euripides, *Bakchai* 1–31). He was also captured by pirates at an early age, who believed he was headed for Egypt, Cyprus, or even the Hyperboreans (Homeric *Hymn to Dionysos*). Herakles, however, was the ultimate traveler, wandering from the eastern Peloponnesos through the world, ending up in northwest Africa (Pliny, *Natural History* 5.1–3) and founding cities everywhere (over 50 eponymous city toponyms are known: see *BA*, index). Because of the connection of both gods to India, Megasthenes (*FGrHist* #715), who wrote an *Indika* in the late fourth century BC, by necessity believed many of the tales of divine presence: this was probably the focus of Eratosthenes' objections, which led him to question the validity of Megasthenes as a whole (see F22). Strabo 15.1.8–9, which immediately follows F21, continues the same theme and may include further comments from Eratosthenes. See also Bosworth, *Commentary* vol. 2, pp. 201–2.

F22. *Summary*: **Deimachos and Megasthenes are not reliable, because they discuss unbelievable things, but Patrokles and others are believable.**

Commentary: Deimachos (or Daimachos) of Plateia (*FGrHist* #716) was sent to the court of Bindusara (Greek Amitrochates, reigned 294–269 BC) at Pataliputra (Greek Palimbothra or Palibothra) in India by the Seleukid Antiochos I: Bindusara was noted for his interest in Greek culture, requesting Greek food and a philosopher but receiving only the first, perhaps brought to him by Deimachos (Athenaios 14.652f–3a), who remains a vague personality. The few fragments of his *Indika* are characterized unfavorably by both Eratosthenes and Strabo: see Klaus Karttunen, *India in Early Greek Literature* (Helsinki 1989) 199. Megasthenes (*FGrHist* #715) is better known: he was stationed at Alexandria Arachosia in the entourage of Sibyrtios, satrap there after 324 BC, and went to the court of Bindusara's father Chandragupta (Greek Sandrakottos, reigned 318–294 BC), also at Pataliputra, probably the first Greek to travel that far east. He too wrote an *Indika*, based on data collected at the court, where he seems to have been a close companion of the king, becoming knowledgeable about the strange (to Greeks) phenomena of India. Although the structure and details of his work are well known (33 fragments survive, some quite lengthy), Eratosthenes was probably the last to see a copy, and despite concerns about reliability, made heavy use of it (F67–76). Since antiquity Megasthenes has been accused of being too credulous, but it must be remembered that he was in a world that was totally bizarre from the Greek point of view, and it was difficult to separate fact from fiction, as modern travelers to India are well aware. The list of anatomically unlikely ethnic groups ("Those Who Sleep in Their Ears," "Without Mouths," "Without Noses," "With One Eye," "With Long Legs," and "With Toes Backward") may in part derive from earlier Greek catalogues of strange peoples: many of them were listed by Skylax of Karyanda (*FGrHist* #709, F7) in the latter sixth century BC and Ktesias (*FGrHist* #688, F45) a century later. The cranes and pygmies of Homer (*Iliad* 3.3–7) were transferred to India, probably based on existing tales of unusually short people there (Ktesias F45, section 21) or a even confusion of ethnyms. The gold-mining ants had been known since Herodotos (3.102–5), one of the most famous Greek anecdotes of Indian lore (see also Nearchos [*FGrHist* #133], F8a): the common explanation is some sort of small mammal. Large snakes are certainly a feature of the region. But Eratosthenes knew that he should be cautious about these tales, and was perhaps overly so. Unfortunately

he was not as suspicious of Patrokles (*FGrHist* #712), who explored the
Caspian region in the early third century BC, but whose reliability was
questionable: see further, F50.

F23. *Summary*: **The Makedonians manipulated mythology in order
to please Alexander: examples include the topography of the tale
of Prometheus and stories about Herakles and Dionysos.**

Commentary: A problem for the Hellenistic geographer (also apparent
in F24) was that Alexander's topographers had tampered with data in
order to enhance his reputation. Even by Eratosthenes' day, a century
later, Alexander's unique vision of topography had become standard,
originating from the official version of his travels by Kleitarchos (*FGrHist*
#137). The issue in question is the location of the cave of Prometheus,
traditionally in the Caucasus, but therefore in a place Alexander had
not been. Kleitarchos, allegedly relying on artifacts conveniently produced by local informants, located the cave in the Parapamisos Mountains (modern Hindu Kush), and then said that this was the Caucasus,
which meant that Alexander had crossed the Caucasus. Eratosthenes
was quite offended at this because it did violence to understanding of
the mountain ranges at the northeast limits of the inhabited world, as
he had data to prove that the Parapamisoi were an extension of the
Tauros (F69) and in no way connected to the Caucasus. In fact, Kleitarchos' interpretation required moving the Caucasus 30,000 stadia
(Strabo 11.5.5). This manipulation of topography in order to connect Alexander with mythology led Eratosthenes vehemently to reject the tales
of Herakles and Dionysos in India, and indeed gave him ammunition to
eliminate data that suggested mythology had influenced geography. For
a detailed discussion of this passage see Bosworth, *Commentary* vol. 2,
pp. 213–19, with the suggestion that it actually belongs in Book 3.

F24. *Summary*: **Those with Alexander manipulated the topography of the regions east of the Black Sea in order to enhance his
reputation, using the evidence of Polykleitos. In doing so, however, they created a number of obvious inconsistencies.**

Commentary: Eratosthenes listed a number of topographical adjustments that were made to enhance the reputation of Alexander, especially
in the region of the Caucasus (see also F23). Alexander seems to have
been obsessed with his failure to reach the Caucasus (he passed well to
the south), probably because it had been known to the Greeks so early

(first mentioned by Aischylos, *Prometheus* 422, 719 [the latter implies that it was well known, however remote]; also Herodotos 3.97). His record keepers altered the topography by connecting the Maiotic Lake (the modern Sea of Azov) and the Caspian Sea, and moving the Tanais River (the modern Don) so that it emptied into the Caspian rather than the Sea of Azov, calling the Iaxartes River (the modern Syr Darya) the Tanais, because it did flow into the Caspian Sea (although its outlet is actually the Aral Sea). Yet the name "Iaxartes" was not entirely superseded, as Arrian recorded that it was the local name for the Tanais. Alexander never came close to the actual Tanais but did reach the Iaxartes in 329 BC (Arrian, *Anabasis* 3.30.6–9), and thus could claim to have reached both the limits of Europe (the Tanais) and the Caucasus (which had to be crossed to come to the river). Certain botanical details were added to prove the point. Much of the manipulation was the work of (or based on data supplied by) a member of Alexander's entourage, Polykleitos of Larisa (*FGrHist* #128), who is little known beyond the geographical data (right or wrong) that he provided Eratosthenes and Strabo. Such geographical revisionism was possible because the topographical understanding of the region was vague and contradictory, and in fact caused a regression in the scant knowledge that did exist. Herodotos (1.203) and Aristotle (*Meteorologika* 2.1) were better informed, knowing that the Caspian was an enclosed sea, but Alexander insisted that it was an arm of the Ocean, which fit into his self-view of having reached the Northern Ocean and eventually being able to use it to reach the Western Mediterranean (Roller, *Pillars* 59–60). The existence of the fir tree, seen as European flora, helped support these views. Portions of the Caspian littoral have a Mediterranean lushness, despite being north of the desolate Iranian plateau: Pliny (31.43) noted the effects of on-shore winds and mountain rain shadows in this region (see Truesdell S. Brown, *Onesicritus* [Berkeley 1949] 90–2). Despite Eratosthenes' strong rejection of these attempts at topographical reordering (and pointing out the contradictions that had resulted, such as the range of the fir tree), he himself fell victim to Alexander's view of topography, continuing to report that the Caspian was part of the Ocean (F110). It was not again determined to be enclosed until the map of al-Idrisi in the twelfth century and the travels of William of Rubruck in the following century. Even then it was still not believed (J. Oliver Thomson, *History of Ancient Geography* [Cambridge 1948] 390). Strabo indicated that there was much more to Eratosthenes' critique of Alexander's topography (see also F23), but unfortunately did not share other examples. This discussion closed Book 1 of Eratosthenes' *Geographika*, a fitting evolution of the theme of geographical fantasies.

Commentary to Book 2

F25. *Summary*: **Mathematics and physics are part of geography. The spherical earth is inhabited all around. The heavens are also spherical.**

Commentary: In the introduction to his second book, Eratosthenes summarized what he had previously published in his *On the Measurement of the Earth*: the diameter of the earth (measured by a meridian) is 252,000 stadia (see infra, pp. 263–7). This calculation is presumably the mathematics and physics to which Strabo referred. Eratosthenes also reemphasized that he believed the earth to be spherical, with habitation not limited to the known areas, ideas new enough to need stressing (see F15). Strabo, 200 years later, felt that Eratosthenes went on at too great length about these issues, perhaps failing to realize how novel Eratosthenes' beliefs had been, and noted that "later writers" (probably including Poseidonios) did not agree with Eratosthenes' methodology.

F26. *Summary*: **Dioptras can be used to measure elevations.**

Commentary: Theon of Alexandria, who lived in the fourth century AC, produced commentaries on mathematical works, especially those of Euclid and Ptolemy, which have no original material but preserve data otherwise lost. He gave a rare insight into Eratosthenes' presumed field methodology, although there is little other evidence that he did fieldwork, and it may be an erroneous reference to Dikaiarchos (see his F120). The *dioptra* was an instrument used for sighting, first discussed in detail by Euclid (*Optics* 19). The altitude of mountains was an early curiosity for Greeks, and it was often believed that mountain summits were so high as to be invisible (Herodotos 4.184), a mythological view reflecting their role as divine residences. It was also thought that the inhabited world was surrounded by mountains, which helped to explain certain visible anomalies about celestial movements (see, for example, the view of Anaximenes of Miletos, which survived as late as Aristotle

[*Meteorologika* 1.13, 2.1]). There seems to have been little effort to determine the specific heights of known mountains until the development of the *dioptra* at the end of the fourth century BC. The first to do this was Dikaiarchos, who calculated the height of Kyllene in Arkadia at 15 stadia (Geminos 17.5). He also determined the height of other mountains, and perhaps wrote a treatise titled *Measurement of Mountains in the Peloponnesos* (Dikaiarchos F2). Eratosthenes may have done some measurement of his own, although none is specified (another measurement of Kyllene at 20 stadia [Strabo 8.8.1] is a possibility). See further, Paul T. Keyser, "The Geographical Work of Dikaiarchos," in *Dicaearchus of Messana* (ed. Fortenbaugh and Schütrumpf, New Brunswick 2001) 353–61.

F27. *Summary*: **A *schoinos* is 40 stadia.**

Commentary: The *schoinos* ("rod" or "rope") was a land measurement, probably Egyptian in origin. Herodotos (2.6) reported that in Egypt it was a distance of 60 stadia, but Strabo (17.1.24) said that it varied from 30 to 120 stadia, even over different parts of the same route. Pliny's calculation of five miles for the *schoinos* used by Eratosthenes results in a length of 9.4 km., but it is unlikely that Eratosthenes was consistent since by his own admission he used distances "handed down" (F131), each source would have had its own *schoinos*, and the stadion itself also varied (see infra, pp. 271–3). The *schoinos* became widely used in the Persian Empire and beyond: the *Parthian Stopping Points* of Isidoros of Charax (*FGrHist* #781, F2), written in the Augustan period, is an itinerary in *schoinoi* from the Euphrates to Alexandria in Arachosia, which demonstrates the longevity of the unit despite several hundred years of Greek rule. Because of the extensive data that were available to Eratosthenes only in *schoinoi*, it was important to establish a conversion between *schoinoi* and stadia, but this did not solve the problem that both units were of varying lengths.

F28. *Summary*: **The circumference of the earth is 252,000 stadia, and its diameter is 42,000 stadia.**

Commentary: Eratosthenes' most famous feat was his calculation of the circumference of the earth, something for which he was remembered long after the *Geographika* was no longer extant. It was presented, along with the methodology, in his *On the Measurement of the Earth* (see infra, pp. 263–7). The circumference of the earth had been a matter of curiosity ever since the Pythagoreans had suggested that it was spherical (supra, p. 5). The earliest extant figure for the circumference is 400,000 stadia, quoted by Aristotle (*On the Heavens* 2.14), but clearly from an earlier source, perhaps Eudoxos of Knidos (J. Oliver Thomson, *History of Ancient Geography* [Cambridge 1948] 116). There is no evidence as to how the figure was attained. A measurement of 300,000 stadia cited by Kleomedes (1.5) must be after 309 BC, since Lysimacheia in Thrace, founded that year (Diodoros 20.29), is part of the calculation, and is normally attributed to Dikaiarchos. Archimedes in his *Sand-Reckoner* has the same figure, implying that it was the standard figure in his day but not his own: in fact, he disagreed with it. Eratosthenes reduced the amount to 250,000 stadia by his famous calculations, but then probably increased it to 252,000 to provide a number divisible by 60, perhaps for ease of handling rather than any conception of degrees (Thomson [supra] 160). With a few exceptions, all sources have 252,000 stadia: Kleomedes (1.7) has 250,000, because he preserved the original calculations that led to that amount; Arrian, in his little-known meteorological work, has the same amount, probably a rounding down by the extant source, Philoponos (*Commentary on Aristotle, Meteorologika* 15.13–15); and Markianos of Herakleia has 259,200 (see F29). Eratosthenes was not the last word on the circumference: Archimedes in his contemporary *Sand-Reckoner* proposed an exceptional 30,000,000 stadia, and Hipparchos added about 26,000 to Eratosthenes' figure (Hipparchos F38: see Dicks, *Hipparchus* 153). Pliny's Roman measurement of 31,500 miles is 46,620 km., with the accepted modern polar circumference about 40,000 km.

Pliny's account closes with the peculiar tale of Dionysodoros of Melos and his visit to the center of the earth. He is not the distinguished mathematician from Kaunos in Anatolia (see Strabo 12.3.16), but nonetheless a geometer of note. The story has the character of a staged event to give folkloric wisdom to arcane mathematical calculations, and thus to provide popular support, perhaps even to supersede Eratosthenes, whose *Measurement of the Earth* would have been unreadable to most. Using the rather rough value of π as 3, his circumference is essentially the same as that of Eratosthenes.

F29. *Summary*: **The circumference of the earth is 259,200 stadia.**

Commentary: Markianos' anomalous figure is unlikely to be a copying error since this number is also divisible by 60. Markianos, of the late Roman period (between AD 200 and 530: see Hans-Armin Gärtner,

"Marcianus" [#1], *BNP* 8 [2006] 304–5), wrote an epitome of the *Periplous* of Menippos of Pergamon, probably of the Augustan period, who may be the ultimate source of the figure. The Dionysios mentioned in the fragment is unidentified, but may be an error for Dionysodoros (F28). There is no obvious reason for the discrepancy.

F30. *Summary*: **The earth is spherical, and the irregularities on its surface do not affect this in any overall sense. There are five zones, with the inhabited portion of the earth a four-sided area in the northern hemisphere, surrounded by the ocean (or, in a few places, uninhabitable land). The part of the northern hemisphere between the equator and the polar parallel is shaped like a spindle whorl, with the inhabited world a chlamys-shaped island within it, 70,000 stadia in length and less than 30,000 in width. The uninhabited part is 126,000 stadia in length and 8,800 in width.**

Commentary: Strabo has outlined Eratosthenes' view of the location of the inhabited part of the earth, as is apparent at 2.5.7 (F34), which follows immediately. F30 is clearer than much of Strabo's text, with propositions and proofs laid out in a straightforward linear style, and thus presumably a direct quotation from Eratosthenes' treatise. Although he is not named, significantly the passage is free of the multiple source citations that pervade most of Strabo's *Geography*. Stressing again that the earth is sphere-shaped (see F25), Eratosthenes deflected the argument that the rough topography of the surface refuted this concept, emphasizing that such irregularities were insignificant. At the same time it was agreed that the earth was not a perfect sphere, neither physically nor mathematically, a deliberate attempt to move away from the Pythagorean and Platonic view of the earth as a harmoniously perfect part of the cosmos.

Next Eratosthenes presumed that the earth had five zones (equatorial, two temperate, and two polar). The division of the earth into latitudinal zones is probably the concept of Eudoxos of Knidos, active in the first half of the fourth century BC, who saw the world as sloping toward the poles and thus used the word κλίμα ("slope": Strabo 9.1.1–2; Polybios 2.16.3, 7.6.1) for the zones. He also began to theorize on how one might use the length of the day for determining latitude. Aratos' *Phainomena*, a poetic version of Eudoxos' treatise of the same name, written in the first half of the third century BC, mentioned the celestial circle (lines 462–510), which, by analogy, could be transferred to the surface of the earth itself, but unlike the celestial circles the terrestrial ones could not be perceived. As expressed by Eudoxos and Aratos the concept of terrestrial circles was still primitive, and as yet there was no good way to determine latitude, although comments in Aratos' poem about the varying length of day point the way. The terrestrial circles were probably further refined by Dikaiarchos, who may have added the arctic circles, almost certainly because of the new evidence provided by Pytheas (Strabo 2.4.2 [= Eratosthenes, F14]; J. Oliver Thomson, *History of Ancient Geography* [Cambridge 1948] 153–5). But the concept of zones and terrestrial circles was still vague even in Eratosthenes' day: his language in the *Measurement* (Geminos 16.6–9) borders on the pedantic in stating repeatedly that the terrestrial circles lie under the celestial ones.

With the five-zoned earth established, Eratosthenes then focused on the northern hemisphere, creating a section of the surface of the earth that defined the inhabited part, again using ideas that probably went back to Eudoxos. Since the External Ocean surrounded the inhabited world, it was like an island. For "inhabited world" Eratosthenes used a relatively new word, οἰκουμένη, which seems to have originated in the fourth century BC to characterize the civilized (i.e. Greek) world as opposed to those not civilized (i.e. the Makedonians) (Demosthenes, *On Halonnesos* 35, *On the Crown* 48). Aristotle (*Meteorologika* 2.5) then refined it to mean the inhabited world, without any ethnic component, in contrast to the parts that were uninhabitable because of cold or heat. As noted, the inhabited world was an island, although Eratosthenes admitted that parts of its perimeter had not been explored and there might be uninhabitable land in these regions, yet this made no difference to the geographer as one could merely join with a line the two termini of coastal exploration. Strabo's unusual lack of sources in this passage (none is mentioned at all throughout the fragment, until a rather weak acknowledgment of Hipparchos [his F36] at the beginning of 2.5.7) is in part because the most recent data on the extremities of the inhabited world were from Pytheas, whom Strabo never believed (F14). Pytheas' voyage to the Arctic extended knowledge of the northwest coast of the inhabited world as far as the inner Baltic, and if he went from that region to the Black Sea by the river network (Strabo 2.4.1 [= Eratosthenes F14]; Pomponius Mela 3.33), this created an eastern connection between north and south (perhaps the source of Eratosthenes' statement that there might be some uninhabitable land beyond the perimeter). Africa had long been circumnavigated (Herodotos 4.42; Roller, *Pillars* 23–7), Alexander's companions Nearchos (*FGrHist* #133) and Onesikratos

(*FGrHist* #134) had defined the Indos-Persian Gulf route, and Megasthenes (*FGrHist* #715, F6c) had supplied data about the coast of India. Patrokles had reported on the sea route from the Caspian Sea to the mouth of the Ganges (Strabo 11.11.6; Pliny, *Natural History* 6.58), although Eratosthenes could not know that such did not exist or that the Caspian was not an inlet of the Ocean. All that remained of the circumference of the inhabited world, it was believed, were two minor stretches: the outer coast of the Arabian Peninsula (probably not known to Greeks until the end of the second century BC: Duane W. Roller, *The World of Juba II and Kleopatra Selene* [London 2003] 231–2) and the coast of the North Sea from the English Channel to the Danish coast, a Roman discovery (Roller, *Pillars* 117–21). Yet these were so brief that they could be ignored.

Eratosthenes likened the inhabited world to a "spindle whorl," cited here by Strabo as if already discussed, although this is his first mention of the word. If Eratosthenes defined his meaning, this was not preserved by Strabo: see further, F31. Another unusual term follows immediately, the "chlamys-shaped" inhabited earth, also used here by Strabo for the first time and without discussion (although a hint of the meaning is provided with the tapering at the ends): see further, F34. Both these words were part of the new vocabulary Eratosthenes adopted but were common enough in Strabo's day to deserve little comment.

The final point to be made is the size of the inhabited world, which Eratosthenes believed to be 70,000 stadia east-west by 30,000 north-south. These figures are probably based originally on Demokritos' idea of the rectangular shape of the inhabited world, with the east-west dimension 1½ times the north-south one. This established the idea that it was not far from the Pillars west to India (first mentioned by Aristotle, *Meteorologika* 2.5), something that obsessed early modern explorers even after the size of the earth was better known. Eudoxos made the east-west length twice the north-south, but Dikaiarchos returned to Demokritos' proportions (Agathemeros 1.2), perhaps 60,000 by 40,000 stadia (Paul T. Keyser, "The Geographical Work of Dikaiarchos," in *Dicaearchus of Messana* [ed. Fortenbaugh and Schütrumpf, New Brunswick 2001] 368). Eratosthenes' figure (70,000 stadia, repeated in F34) is actually less than a totaling of the distances cited (see F37), a typical example of the parts-and-sum problem that had plagued Greek writers since Herodotos. A north-south distance of 30,000 stadia (F34) is also short of the sum of actual measurements. Eratosthenes stressed that the inhabited world was limited on both its north and south sides because of an inhospitable climate, an early idea best expressed in the Hippokratic *Airs, Waters, and Places* of the latter fifth century BC, which set forth how those living at the extremities of the inhabited portion, especially the far north, were likely to have peculiarities and failings of character. Although this theory was breaking down in Eratosthenes' day, with Pytheas' voyage and the Ptolemaic explorations into the tropics, it was still generally accepted. Eratosthenes also believed that it was nearly twice as far around the uninhabited portion (from the Pillars west to India) than in the other direction, providing an indirect figure of 196,000 stadia as a circumference on the latitude of the Pillars, but further analysis is difficult because of textual problems at this point in Strabo's *Geography* (see Strabo, *Géographie* 1.2 [ed. Germaine Aujac, Paris 2003] 158–9).

F31. *Summary*: **The issue of whether the spindle whorl is inhabited outside the known portions is not part of the discipline of geography, but if it is inhabited those living there would not be like us, but in another inhabited world.**

Commentary: Eratosthenes used the term σπονδύλος ("spindle whorl") to describe the shape of the section of the earth's surface that contained the inhabited earth, an area bordered by two parallels. The term originally meant a vertebra (Plato, *Timaios* 74a) and has a consistent medical and zoological usage, but Eratosthenes was relying on a domestic definition that also existed in Classical Greek, perhaps inspired by the vivid description of the great cosmic spindle whorl in Plato's *Republic* (10.14). The word was common for many things shaped like a vertebra (such as various flora and architectural details: see *LSJ*), so Eratosthenes' use of a familiar Greek word assisted in making his researches more palatable. The spindle whorl, then, is the entire circumference of the earth between two parallels, which Eratosthenes cut in half because the inhabited world was in only one hemisphere. Since the half spindle whorl was only part of the surface of the earth, the question arose as to whether the other half, or indeed other potential spindle whorls, could include inhabited worlds. Eratosthenes briefly noted the possibility (admitting that their inhabitants would be unlike those in the known world), an idea going back to Plato's *Atlantis* but having more recent authority in Epikourean thought (*Letter to Herodotos* 74; see also Lucretius 2.1074–6), but then dropped the matter, commenting that such speculation was not geography.

F32. *Summary*: **The length of the world is more than twice its width.**

Commentary: The source of Agathemeros' list of those who considered the inhabited world longer east-west than north-south is not obvious. At some date in the Roman period he wrote a summary of geography that opens with a brief account of the history of geography from Anaximandros to Poseidonios (1.1–2). Included is the comment that Eratosthenes believed that the inhabited world was more than twice as long east-west as north-south (implicit in F30), but Agathemeros showed little actual familiarity with Eratosthenes, as he described the inhabited world in terms of various shapes but did not mention Eratosthenes' striking use of the chlamys image (F34). It is probable that the account comes from a catalogue of the history of geography, rather than Eratosthenes himself. For Demokritos and geography, see F1; the significance of Eudoxos and Dikaiarchos is discussed in F30. On Agathemeros generally see Aubrey Diller, "Agathemerus, *Sketch of Geography*," *GRBS* 16 (1975) 59–76.

F33. *Summary*: **It is perfectly natural to have the inhabited world longer from east to west, and, were it not for the size of the Atlantic Ocean, one could sail from Iberia to India along the same parallel, a distance of less than 200,000 stadia. Much has been written about how to divide continents, whether it should be done by rivers or isthmuses, but there are often no exact boundaries between them. Greeks began to conceive of continents when they related their own world to Karia, and, eventually, as their perspective widened, they saw the three continents of Europe, Asia, and Libya.**

Commentary: In a long and difficult passage that can border on the incomprehensible, Strabo examined the division of the inhabited world into continents. Although it seems to include one of the few direct quotations of Eratosthenes, the text is Strabo at his most inscrutable. The first section repeats some of the material from F30–2 and may be close to Eratosthenes' original text. Eratosthenes seems to have been a little uncertain about some of his assumptions, stressing how normal it was to consider the inhabited world to be longer east-west than north-south, despite the fact that this had been believed for 200 years (see F30). He also emphasized that his view was predicated on a spherical earth (see F15, 25). The idea that one could sail west to India was an obvious conclusion resulting from the idea of the spherical earth, first suggested by Aristotle (*Meteorologika* 2.5; *On the Heavens* 2.14), although he acknowledged that the distance on the open sea made it unfeasible. For Eratosthenes' figure of about 200,000 stadia for the journey, see F30. Strabo made it clear that Eratosthenes was using the latitude of Athens for this distance, which was close to the latitude of the Pillars (the difference is only 2°), and it hardly mattered that the latitude of Athens is far north of India (it is, in fact, north of that of Tibet). These comments demonstrate the rudimentary construction of a main terrestrial parallel, whatever its inaccuracies (see F37–8). But Strabo also raised the question whether there might be a continent intervening between the Pillars and India, since the parallel was through the temperate zone, the most inhabited. Eratosthenes dismissed such speculation as outside the discipline of geography (F30). Strabo accused Eratosthenes of pedantic argument and could not resist another forced comment about his treatment of Homer (F2–13), again showing that Strabo did not truly perceive the novelty of these views in Eratosthenes' day (F25).

The next subject is the continents. Hekataios of Miletos was probably the first to conceive of them, believing that there were two, Europe and Asia (see *FGrHist* #1 and Hippokrates, *Airs, Waters, and Places* 13). It is by no means clear whether he had any theoretical structure behind this division: Strabo provided an engaging tale about Greeks coming to the Anatolian coast and perceiving that Karia was somehow different from the Greek world (carefully pointing out that this was before Greeks lived in Anatolia), and as Greeks traveled more and more this sense of difference also expanded to other areas. This is more folk wisdom than geographical theory, but hints at an original ethnic basis for continental theory, remindful of the original use of οἰκουμένη to distinguish Greeks and non-Greeks (see commentary to F30). At some early date the two continents became three with the addition of Libya; Herodotos (4.42) took this number for granted.

Once the three continents were established, there was no further argument as to their existence or location, and no new continents were to be added for 2,000 years. The difficulty, however, was how to divide them, something none too clear. The problems were the separation of Asia and Libya and the eastern parts of the boundary between Europe and Asia (the Hellespont, Bosporos, and Black Sea provided an easy division in the west). Eratosthenes devoted some detail to the question of continental boundaries, reporting that there were several ways to distinguish them, especially using rivers and isthmoi. The Phasis River at the eastern end of the Black Sea was considered by some to be the boundary between eastern Europe and Asia, although Eratosthenes seems to have preferred the Tanais, which had also been accepted in the fifth century BC (Herodotos 4.45, who made it clear that both divisions were before his time). But the real problem was the distinction between Asia and Libya, and what to do about the Nile. In many ways it seemed an ideal boundary, except that the same peoples lived on either side of it. Herodotos was disturbed at using it for a boundary, which indicates that the idea predated him, although he never specifically stated any alternative, such as the isthmus to the east. Eratosthenes on the other hand seems to have preferred the Nile, but no discussion is preserved, and the Nile as boundary, whatever the ethnic repercussions, lasted well into Roman times, with its right bank and the strip to the Red Sea considered part of Asia (see Pliny, *Natural History* 5.52–4, 6.177). Rather than discuss the continental boundaries more fully, Eratosthenes (at least as Strabo presented him) moved to a more philosophical examination of the very nature of geographical boundaries.

Eratosthenes stressed that boundaries were generally invisible, using two Attic demes as an example. Melite lay to the west of the Agora: he may have chosen it because it was one of the more famous localities in the city, with a long list of notable residents from Themistokles to Epikouros. Kolyttos, to its south, was somewhat more obscure, although allegedly the home of the misanthrope Timon (Loukianos, *Timon the Misanthrope* 7; for the demes, see John S. Traill, *The Political Organization of Attica* [*Hesperia Supplement* 14, 1975] 40, 50). Those who knew Athens would be familiar with these urban districts, and it could be easily demonstrated that there was no marked boundary between them as one walked from one to the other, yet each was unique. Two further examples reinforced this point of view. The plain of Thyrea lay on the west side of the Gulf of Argos, almost equidistant between Argos and Sparta, and had been constantly in dispute between the two Greek states until a settlement was imposed by Philip II of Makedonia (Pausanias 2.38.5; Yves Lafond, "Cynuria" [#1], *BNP* 3 [2003] 1063). Oropos had a similar fate, topographically Boiotian but close to Athens, and containing the famous Amphiaraion. Eratosthenes stressed that such localities were often disputed, but that this really had nothing to do with continental boundaries except for those deliberately seeking some point of contention. As usual, Strabo tended to obscure Eratosthenes' argument, and objected to it, emphasizing that political disputes were just as likely to happen at the continental level as at the local one. As an example Strabo mentioned Egypt, aware of the recent history of the territory, but at the end of the fragment he seems to have returned to Eratosthenes, arguing that it is unimportant to define every last point that divides two continents. The statement that the continents are not islands (made twice), but merely part of the island of the inhabited world, reflects some unknown early theory.

F34. *Summary*: **One-fourth of the length of the equator, 63,000 stadia, is the distance from the equator to the pole. From the equator to the summer tropic is four-sixtieths of the length of the equator. All this is calculated from known measurements. The summer tropic is the parallel through Syene, and the tropic is there because there is no shadow in the middle of the day. The meridian of Syene extends along the Nile from Meroë to Alexandria; from Syene to Meroë is 5,000 stadia, and 3,000 farther south (the parallel through the Kinnamomophoroi region) it is not inhabitable because of the heat. North of Alexandria, the meridian extends to Rhodes and eventually Borysthenes, beyond which it is uninhabitable. The farthest north region is Thoule, according to Pytheas. The entire north-south width of the inhabited world is about 30,000 stadia, and its length (Iberia to India) about 70,000. This creates a chlamys shape for the inhabited world.**

Commentary: Having established the polar circumference at 252,000 stadia, Eratosthenes divided it for convenience into sixtieths. The use of 360° to a circle was not yet in use: it first appears in the *On Rising Times* of Hypsikles of Alexandria, written in the early second century BC (see Dicks, *Hipparchus* 148–9) and was first used extensively by Hipparchos later in the century. Eratosthenes established the distance from the equator to the pole as fifteen-sixtieths or 63,000 stadia, and four-sixtieths (16,800 stadia) from the equator to the summer tropic, establishing his first parallel at that point, through Syene in Upper Egypt, stressing the importance of Syene's location (see further, F40–3). The meridian of Syene was considered to be the same as that of Alexandria, although the latter lies nearly 150 km. to the west. Pelousion would have been better, but using Alexandria allowed Eratosthenes to make use of the existing surveyed distances along the Nile (about 5,000 stadia), as well as sailing distances north of Alexandria. He extended the parallel up the Nile to Meroë, a lucky choice, since despite the twisting and turning of the river Syene and Meroë were close to being on the same meridian. Meroë had been known to Greeks since the time of Herodotos (2.29) as a major city at the fringes of the world, so was particularly suitable for defining the southern limits of habitation. A series of little-known explorers had been sent by Ptolemaios II to the city and beyond (Pliny, *Natural History* 6.175–95), providing a recent body of data for Eratosthenes to use (see F98; the most complete Greek description of the city is Diodoros 1.33). It was also convenient that Meroë was said to be as far south of Syene as Alexandria was north (actually Meroë is over 100 km. closer but this is the airline distance), all creating a symmetry that allowed Eratosthenes to begin to lay out his conception of the world. But Meroë was in fact not the end of habitation, because 3,000 stadia beyond was the parallel of the land of the Kinnamomophoroi. This would be a point somewhat south of Khartoum, but Eratosthenes' data are quite vague. In ancient times κιννάμωνον was used indiscriminately for several closely related aromatics (rarely used for food), cinnamon, cinnamomum, and cassia, often with little precise distinction between the three, a confusion that has lasted into modern times. They came from the Far East, even beyond India, to the mouth of the Red Sea, hence the ethnym, for the Greeks used the term "Land of the Kinnamomon Bearers" for the region where they first became aware of the plants, although none of them was grown there. Herodotos (3.111) was the earliest to cite these aromatics, but by Hellenistic times they were regular imports to the Mediterranean (Andrew Dalby, *Food in the Ancient World from A to Z* [London 2003] 87–8; Lionel Casson, *The Periplus Maris Erythraei* [Princeton 1989] 122–4). The Land of the Kinnamomon Bearers (the modern Somali coast) was actually far to the southeast (over 1,000 km.) of the point Eratosthenes defined on the upper Nile, but, as essentially the only known toponym south of Meroë, it served as a convenient southernmost parallel. It was said to be too hot for anyone to live farther south: the reports of those who had circumnavigated Africa had provided no data to dispute this. Thus Eratosthenes was able to establish three parallels along his prime meridian and to compute the total distance from the Kinnamomon Bearers parallel to Alexandria as 13,000 stadia. These measurements were obtained from overland routes, but it all fell into place with his calculation of the circumference, and he could note that it was a further 8,800 stadia to the equator.

Eratosthenes then extended the meridian north from Alexandria to Rhodes and eventually to the Borysthenes region on the north shore of the Black Sea, the collection of Milesian settlements at the mouth of the Borysthenes (modern Dnieper) River. This portion of the meridian made a great arc from Alexandria to Lysimacheia and back to Borysthenes, showing that it was obtained from sea routes, although still an improvement on previous meridian calculations. Herodotos, perhaps the first to consider the topic, created a meridian that went north across Anatolia but then veered sharply northwest from Sinope across the Black Sea to the mouth of the Istros (2.34). There was an essential contradiction in attempting to create straight lines (whether meridians or parallels) on the surface of the earth by only using known points, something that Hipparchos (his F12, 18) sharply criticized.

When Eratosthenes extended his prime meridian north to the Borysthenes, the question arose of what was farther north. One could not dismiss it as uninhabitable, since it had long been known that the Skythians lived north of the Black Sea. Strabo mentioned two ethnic groups, the Roxolanoi, who had warred against Mithradates the Great (7.3.17), and the Sauromatai, but neither seems to be known before Roman times, so cannot be part of Eratosthenes' text. Yet Strabo's citation of the "north of Brettanike" leads into the major source for the far north, Pytheas of Massalia. At section 8 Strabo shifted from the common spelling "Brettanike" to the rare "Prettanike" (a detail ignored or even "corrected" by some editors), which seems the original spelling of the toponym, with the more familiar "B" spelling only coming into use in Roman times (see the Roseman edition of Pytheas, p. 45). This confirms that Strabo has changed sources, ceasing his editorial comments and returning to Eratosthenes, who was quoting Pytheas directly. Of course any mention of Pytheas invoked Strabo's ire (see F14), and the following passage has brief comments from Pytheas (via Eratosthenes) with lengthy and anachronistic interpolations by Strabo. Eratosthenes learned several points from Pytheas. First, Thoule was the northernmost of all places, near the arctic circle, which had the same relationship to the north pole as the summer tropic had to the equator. This meant that the inhabited world and prime meridian could be extended far to the north (see F35). Second, Byzantion and Massalia were on the same parallel, so Pytheas' travels north of Massalia could be connected to the prime meridian. Unfortunately this is an error of over 2° of latitude (Massalia is farther north) that was never rectified in antiquity (Ptolemy, *Geography* 2.10.8, 3.11.5). The source of this problem is not clear—either Pytheas himself or Hipparchos' interpretation of the raw data (Dicks, *Hipparchus* 182–3)—yet any error is not as important as the ability to tie Pytheas' latitude calculations in the north to the eastern Mediterranean, allowing the conclusion that the British Isles are in the latitude of the Borysthenes region (although the error has increased: it is at the latitude of central France), and placement of Thoule (F35) within the overall scheme. At this point Eratosthenes merely added 4,000 stadia to reach unspecified "northern regions," perhaps indicating his own uncertainty about Pytheas' data.

Eratosthenes then referred back to the quadrilateral he had constructed in F30 to demonstrate that now he had the data to create a north-south measurement of the inhabited world at 30,000 stadia, coupled with his east-west measurement of 70,000 (see F37), which would all fit within the quadrilateral. He introduced another of his descriptive terms (hinted at in F30), saying that the inhabited world was shaped like a chlamys (χλαμυδοειδής). This word is not documented earlier, and is another use of domestic imagery to make his views more palatable. One suspects it has a vernacular history not recorded in literature. The chlamys was an outer garment, perhaps originally worn by horsemen (Xenophon, *Anabasis* 7.4.4), seemingly unknown before Classical times. It was common in art, most notably on the riders on the Parthenon frieze. Eratosthenes' only concern was its shape, essentially a rectangle with rounded or somewhat tapered corners, the perceived narrowing of the inhabited world at its extremities (India and Iberia).

F35. *Summary*: **The width of the inhabited world, from the Kinnamomophoroi (or Taprobane) to Thoule is 38,000 stadia.**

Commentary: The fragment summarizes the prime meridian of Eratosthenes (constructed, so some extent, in F34), but with slightly different figures and terminology. The Hellespont is used rather than Lysimacheia, a minor change since the two locales are only 10 km. apart, but suggesting that Eratosthenes may have had some data from before the founding of Lysimacheia in 309 BC. There is a specific distance for Thoule, 11,500 stadia north of the Bosporos, and additional details from Pytheas regarding the location of Thoule and the existence of the frozen sea (which helps identify Thoule as Iceland: see Roller, *Pillars* 80–1). The meaning of the frozen sea has long been argued: it is probably not the polar pack ice (which was generally much farther north) but drift ice or the great glacial inlets of southeastern Iceland. At the other extremity of the world are two toponyms not previously mentioned: the Egyptian Island and Taprobane (for which see F74). The Egyptian Island was a refuge of a group of soldiers who revolted from King Psammitichos (II?) in the early sixth century BC: the tale was told by Herodotos (2.30–1), who placed it two months upriver from Meroë, a more reasonable southern limit of the world than the Land of the Kinnamomon Bearers, far to the southeast (see F34).

This fragment appears early in Strabo's *Geography*, well before the more detailed discussion of F34. It is probably from a late summary of

Eratosthenes (it uses the form "Brettanike," which demonstrates it is not from him directly: see F34), but no source is cited.

F36. *Summary*: **The mouth of the Borysthenes is somewhat over 23,000 stadia from Meroë.**

Commentary: The distances here differ slightly from F34–5, although close to those of F30 (a mere difference of 100 stadia from Meroë to the Hellespont), and probably come from the same source, with the slight error due to sloppiness on the part of Strabo.

F37. *Summary*: **The length of the inhabited world, from India to the Pillars of Herakles, is 70,800 stadia, with the bulge of Europe beyond the Pillars an additional 3,000 stadia. There are even more promontories beyond this, as far as Ouxisame in Keltic territory, according to Pytheas.**

Commentary: Eratosthenes had stretched the east-west length of the inhabited world to more than twice the north-south, refining Eudoxos' measurement of exactly twice (see F30). Strabo objected to this, largely because the elongation was to account for Pytheas' data from the far northwest, but the criticism was valid in that it created an inhabited world that was astonishingly long. Totaling Eratosthenes' measurements from India west, Strabo showed that his dimension was nearly 74,000 stadia (not the 70,000 of F30 and 34). By any calculation of the stadion this is excessive, since the actual distance from the mouth of the Ganges to the latitude of the southwestern point of Europe (Cabo de San Vicente, "the bulge of Europe") is about 9,500 km. Although the uncertainties about the length of the stadion makes any conversion difficult, it is an error of perhaps over 2,000 km. (J. Oliver Thomson, *History of Ancient Geography* [Cambridge 1948] 165), one that lasted into modern times and affected Renaissance exploration, since it made the distance west to India seem shorter than it was. A brief itinerary up the Atlantic coast of Europe (obtained from Pytheas) provides the reason for this elongation: beyond "the bulge of Europe" (see F53, 153) are three toponyms associated with the far northwest corner of France. The Ostimioi were a local ethnic group (known to Caesar as the Osismi, *Gallic War* 2.34, 3.9, 7.75). Kabaion may be a general term for the entire west end of Brittany or specifically Pointe du Raz. Ouxisame seems to survive in modern Ushant, the island off the northwest end of France, which was actually a peninsula in antiquity. These equations are rather speculative (see the Roseman edition of Pytheas, 39, 122–4), although their general location is assured, and Strabo was correct to question their effect on the total east-west length, since Ushant is well east of the latitude of the Portuguese coast. There are frustrating hints of how much is lost from Pytheas' treatise, with hints of other ethnyms and the tantalizing three-day sail to Ouxisame, which would be enlightening about his route. For Ierne, see F53. The final sentence of the fragment remains enigmatic, since even without the addition of 4,000 stadia the east-west dimension would be more than twice the north-south.

F38. *Summary*: **The inhabited world is 83,800 stadia in length and 35,000 in width. The Euphrates and Tigris are 3,000 stadia apart.**

Commentary: The treatise *Measurement of the Entire Inhabited Earth*, part of the corpus of the *Minor Greek Geographers*, is a brief geographical summary, of uncertain date and of little interest except for the unique citation of Eratosthenes, which provides dimensions of the inhabited earth larger than anywhere else. The tone and some of the toponyms in the work suggest a late antique date, and the only manuscript is fourteenth or fifteenth century. It is difficult to believe it represents a reliable tradition from antiquity (see Aubrey Diller, *The Tradition of the Minor Greek Geographers* [n.p. 1952] 39–40).

F39. *Summary*: **The inhabited world is an island, and the Ocean is found in all directions in one continuous sea. Most of its circumference has been sailed, and where it has not been, it is because of tidal conditions, not the obstruction of land masses.**

Commentary: F39 seems close to Eratosthenes' original text, given its straightforward manner. Stressing that the inhabited world is an island (see F30), he emphasized this means that the External Ocean must be continuous, an idea that essentially went back to Homer and Hesiod (*Odyssey* 12.1; *Works and Days* 566), but which had come into question because of reports of ships being unable to go beyond a certain point. Eratosthenes was thinking especially of the Persian Sataspes, in the early fifth century BC, who was sent to circumnavigate Africa and turned back somewhere in West Africa because his ship became stuck (Herodotos 4.43), but also because he was frightened by the isolation (ἐρημία) that he encountered, a word repeated in Eratosthenes' account.

In the following century a Greek traveler in the same region reported that one could not sail beyond the Carthaginian trading post of Kerne (Pseudo-Skylax 112). Although these difficulties can be attributed to the adverse currents and tides, they were often explained by having encountered an otherwise-unknown land mass, another point not missed by Renaissance explorers. In Eratosthenes' day, some believed that the Indian Ocean was an enclosed sea, and even as late as Hipparchos (his F4) the continuous nature of the External Ocean was still questioned. Eratosthenes was unconcerned that the inhabited world had not been completely circumnavigated, optimistically stating that most of the perimeter had been covered and that one could presume its island nature. Yet theories about other continents (F31) obscured the question about the nature of the External Ocean. He realized that tides were part of the difficulties sailors faced but had little understanding of them (F16), as no study of the tides had yet been written.

F40. *Summary*: **Philon recorded that at Syene the sun is at the zenith 45 days before the summer solstice.**

Commentary: Philon (*FGrHist* #670) was a Ptolemaic officer ("Philone praefecto," Pliny, *Natural History* 37.108) who explored the upper Nile and the Red Sea early in the reign of Ptolemaios II. His report was titled *Aithiopika* (Antigonos of Karystos, *Paradoxes* 145), but he is only known from the material in F40 and the two citations above. Nevertheless his treatise included detailed scientific observation, as he was the first to report the latitude of Meroë, having spent some time there. He took a number of measurements and may have provided Eratosthenes with the distance from Meroë to Alexandria (about 10,000 stadia, F34), whether by overland survey or astronomical calculation. For the sun to be at the zenith 45 days before the solstice places Meroë at latitude 16°46' north, an accurate measurement (Dicks, *Hipparchus* 128).

F41. *Summary*: **At Syene there are no shadows for 45 days before and after the summer solstice. Similar conditions occur at Berenike.**

Commentary: This is the most complete statement about the phenomena in the region around Syene (see also F40, 42), citing several localities where there are no shadows (all lying between 18° and 24° north). The citation from Onesikratos (*FGrHist* #134), one of Alexander's companions, regarding similar phenomena in the vicinity of the Hypasis

River (which lies between 24° and 32° north latitude) is strangely inappropriate, and perhaps contributed to the misplacing of India that pervades Eratosthenes' treatise. A thorough description of Syene by Strabo (17.1.48) does not mention Eratosthenes. It seems unlikely that the famous well was constructed merely for observation, but it may have become a tourist site by Pliny's day. Berenike, on the Red Sea coast, generally known as Berenike Trogodytika, founded by Ptolemaios II and named after his mother (Getzel M. Cohen, *The Hellenistic Settlements in Syria, the Red Sea Basin, and North Africa* [Berkeley 2006] 320–5), is only about 20 km. south of the latitude of Syene. Ptolemais to the south, on the coast between 18° and 19° north latitude, was founded at roughly the same time by a certain Eumedes as a hunting outpost (Strabo 16.4.7; Pliny, *Natural History* 6.171); see further, Lionel Casson, *The Periplus Maris Erythrae* (Princeton 1989) 100–1; Cohen (supra) 341–3. The emphasis on shadows falling to the south (also in F42 and 43), something possible shortly after one crosses the tropic, demonstrates how unusual the phenomena of the area seemed to Greeks (see also Herodotos 4.42). Trogodytika is the coastal region of the Red Sea, often misspelled in both ancient and modern times as "Troglodytika," yet this is not the Land of the Cave Dwellers but an indigenous term. The Trogodytes had been known to Greeks since the fifth century BC (Herodotos 4.183) and were under loose Ptolemaic control from the time of Ptolemaios II on, noted as agriculturalists and elephant hunters.

F42. *Summary*: **At Syene a well was dug that showed the lack of shadows at the summer solstice. This also happens at other places, and in part of the year the shadows fall to the south.**

Commentary: Pliny, perhaps using Philon (although not cited by name here: see *Natural History* 37.108), repeated some of the data from F40 and 41 about the midsummer sun and its lack of shadows at Meroë. For Berenike and Ptolemais see F41. Pliny's dynamic statement about Eratosthenes recognizes the significance of this region to his work.

F43. *Summary*: **At Syene there are no shadows at the summer solstice. At Meroë shadows fall to the south 90 days of the year, and the locals are called the Antiskioi.**

Commentary: Ammianus Marcellinus' description of Syene is more anecdotal than the scientific data in F41–2, and does not mention Eratosthenes, but continues the obsession with shadows falling to the south. It also provides the chauvinistic Greek ethnym "Antiskioi" ("Those Whose Shadows are Opposite"), a word not documented before late Roman times and probably not used by Eratosthenes.

F44. *Summary*: **The sphere-shaped earth has five zones: the two chilled ones at the poles, then the two temperate ones, and the single burned zone that is divided by the equator. The northern temperate zone is where we live.**

Commentary: Eratosthenes' discussion of the terrestrial zones may only have appeared in his *Measurement* with perhaps a brief summary in Book 2 of the *Geography* (see F30). The theory of zones seems to have been developed by Eudoxos, refined by Dikaiarchos, and finalized by Eratosthenes, who may have added the two arctic ones because of data from Pytheas (see commentary to F30). The figures given by Geminos do not correspond to those used elsewhere by Eratosthenes, so there are probably intervening sources. There was some disagreement as to whether the middle ("burned") zone was one or two, with one on each side of the equator: see further, F45.

F45. *Summary*: **The area under the celestial equator is temperate**.

Commentary: Although it was standard opinion that the equatorial regions were uninhabitable because of heat (F34), evidence was building to suggest the contrary. The early circumnavigation of Africa may have provided the first hints, but the Ptolemaic explorers who spent time at Meroë, such as Philon (F40) and Dalion (*FGrHist* #666), who went far beyond Meroë (Pliny, *Natural History* 6.183), probably gained more information. It is difficult to believe that Simonides (*FGrHist* #669), who spent five years at Meroë, perhaps as an ambassador of Ptolemaios II, did not learn about what was to the south. It had also been known for some time that there were high mountains at tropical latitudes on the coast of West Africa (Hanno seems to have reached Mt. Cameroon [Roller, *Pillars* 39–41]), and it may have been believed that these extended across the continent, much like the east-west ranges of Europe and Asia. So Eratosthenes, while generally holding to the theory of uninhabitable equatorial regions, did suggest that there might be a narrow temperate region right at the equator, although the formalizing of it into a seventh zone (presuming two burned zones: see F44) was probably a later concept (Walbank, *Commentary* vol. 3, pp. 575–6, but see Aubrey Diller, "Geographical Latitudes in Eratosthenes, Hipparchus and Poseidonius," *Klio* 27 [1934] 263–5). It remained for Polybios to write a treatise on the equatorial regions (which he had visited in West Africa), formally stating that they were at high altitude (Geminos 16.32 [= Polybios 34.1.7]; Kidd, *Commentary* 236–8). Somewhat later, perhaps not until around AD 100, an explorer named Diogenes learned, either by autopsy or hearsay, about the mountains and lakes of central Africa where the Nile originated (Ptolemy, *Geography* 1.9, 17; 4.6–8).

Commentary to Book 3

F46. *Summary*: **The inhabited world can be defined by a series of parallels cutting across the meridians.**

Commentary: Having laid out some details of the extent of the inhabited world in Book 2, Eratosthenes now turned to its specifics. Despite the statement at the beginning of F47, F46 is probably the opening to Book 3. Without providing measurements or topographical details, he explained his methodology of comprehending the inhabited world. Building on the work of Dikaiarchos, who created a prime meridian and a prime parallel, Eratosthenes devised a series of both, so that the entire inhabited world would be covered with a grid of parallels and meridians. As far as can be determined, no one had ever done this before.

F47. *Summary*: **The inhabited world is divided into two parts by a line from the Pillars of Herakles to the mountains north of India. Southern India is on the parallel of Meroë, 15,000 stadia south of the mountains according to Patrokles. From the Issic Gulf to the Pontos is 3,000 stadia, and from Meroë to the Hellespont 18,000, the same as from southern India to Baktra.**

Commentary: Eratosthenes' baselines had been hinted at in Book 2 (F34, 37), but it was now made clear that his east-west baseline, his prime parallel, would divide the inhabited world into a northern and southern part. Dikaiarchos had devised a main terrestrial parallel from the Pillars through Sardinia, Sicily, the Peloponnesos, southern Anatolia, and to the Imaos Mountains (his F123 [= Agathemeros 1.5]). Using the Imaos (essentially the Himalayas) as his eastern terminus demonstrates access to material published by the companions of Alexander. The parallel expectedly wobbles, although not badly, but inclusion of Sardinia, far out of line, is probably an error somewhere between Dikaiarchos and Agathemeros. Eratosthenes altered this baseline somewhat, evidently eliminating Sardinia and running it through the Straits of Messina, southern Attika, and Rhodes to the Gulf of Issos, the east-ernmost point of the Mediterranean, somewhat south of Dikaiarchos' line. Eratosthenes' line also wobbles, but not as much: Rhodes and the Pillars are almost exactly on the same latitude, but the Straits of Messina and southern Attika are nearly 2° to the north. A straight line from the Pillars to Rhodes would have passed south of both Sicily and the Peloponnesos, and since Eratosthenes was dependent on sailors' reports, he may have felt the need to stay close to the shipping lanes to avoid a line totally in the open sea. Rhodes had become far more important since the time of Dikaiarchos, which may have led Eratosthenes to make it the point where his two prime baselines crossed (see F34).

From the Gulf of Issos east the parallel was somewhat more hypo-thetical, something Eratosthenes realized. The Tauros Mountains in southern Anatolia were believed to be the western end of a great range that ran all the way to the Imaos north of India (the Tauros, El-burz, Hindu Kush, and Himalayas), a generally accurate theory. The mountains lie between 35° and 37° north latitude as far as the Hindu Kush, but then across northern India they angle in a great arc to the southeast, through 10° of latitude. Yet some early plan available to Er-atosthenes showed exactly the opposite, with the eastern end of the mountains turning toward the north. Eratosthenes was sensitive enough to realize the error, but corrected it only halfway, making the Imaos run east-west rather than to the northeast as on the old plan, or to the southeast as they actually do. Believing that the south end of India was the same latitude as Meroë (a significant feat, perhaps the first example of connecting points so far away by latitude), and relying on the report of Patrokles (see below) that India extended 15,000 sta-dia from north to south, he was able to place India on the map.

As is often the case, the profundity of the conclusions overshadows the errors in measurement. In addition to the problem of the orienta-tion of the Imaos, Cape Comorin, the southernmost point of India, is in fact 9° south of Meroë, and it is difficult to determine where the north end of Patrokles' 15,000 stadia should be, because of the angling of the mountains (the Himalayas can be from 2,400 to over 3,000 km. distant from Cape Comorin). Moreover 15,000 stadia seems far too little, result-ing in an India about two-thirds its actual size with the southern third of the subcontinent missing.

Patrokles (*FGrHist* #712) was a Seleukid officer stationed in the east during the reigns of Antiochos I and Seleukos I (Pliny, *Natural His-tory* 6.58), which would make his activities between 312 and 261 BC. He was sent on an extensive topographical mission east of the Caspian Sea,

the very region where Alexander's companions had manipulated the topography (F23–4), so the Seleukid government may have felt the need
for accurate information, especially about the routes along the eastern
edge of the Seleukid territories. Occasionally the terse style of an official report is preserved by Strabo (the only source for the fragments of
Patrokles), as at 2.1.15, probably passing from Patrokles to Strabo via
Eratosthenes. See also W. W. Tarn, *The Greeks in Bactria and India* (revised third edition, ed. Frank Lee Holt and M. C. J. Miller, Chicago 1997)
488–90; John R. Gardiner-Garden, *Greek Conceptions on Inner Asian
Geography and Ethnography from Ephoros to Eratosthenes* (Bloomington, Ind., 1987) 39–44; Klaus Karttunen, *India and the Hellenistic World*
(Helsinki 1997) 257. Unfortunately Patrokles seems to have relied too
much on hearsay, suggesting that one could sail from the Caspian Sea
to India (his F4 [= Strabo 2.1.17]) and not obtaining accurate information about the rivers of central Asia (his F5–6 [= Strabo 11.7.3, 11.11.5]).
Nevertheless Eratosthenes generally considered him reliable and he
did contribute to Greek knowledge of the region.

Eratosthenes also recorded a north-south width of 3,000 stadia
across Anatolia, from the northeast corner of the Mediterranean to the
Black Sea, the same as a figure that he had for the width of the mountains. He then envisioned a line from the north coast of Anatolia due
east, considering it the northern rim of the mountains (although it would
pass through the middle of the Caspian Sea). His diction implies that he
was following trade routes, especially east of the Caspian. The west end
of the line could be extended to the Hellespont, which Eratosthenes had
already calculated was about 18,000 stadia north of Meroë (F35), so this
allowed him to suggest that this figure corresponded with Patrokles'
width of India (15,000) and the width of the mountains (3,000), making
the correspondence between Meroë and the south end of India. Unfortunately it placed India about 700 km. too far to the north.

F48. *Summary*: **The Kaspian Gates are the boundary between the
northern and southern parts of the inhabited world, and Media
and Armenia are in the southern portion.**

Commentary: Eratosthenes used the Tauros range to divide the inhabited world into northern and southern portions (F47), and placed Armenia and Media in the southern part. Since both territories straddle the
mountains, it would be difficult to make the distinction, one of the traps
Eratosthenes could fall into while structuring the world. Strabo, somewhat better informed about this region (Armenia was a Romanized

kingdom in the Augustan period), could quibble about location. The Kaspian Gates are the primary means of access from the plateau of Media to the littoral of the Caspian Sea, the route taken by Alexander (Arrian, *Anabasis* 3.20) and seemingly unknown to Greeks before that time. Even in Strabo's day (11.12.1) they were seen as the end of the civilized world. It is not certain which of the several passes through the modern Elburz is meant: the most likely is the Sar Darreh, east of Tehran (J. F. Standish, "The Caspian Gates," *G&R* 2nd ser. 17 [1970] 17–24). The gates were cited repeatedly by Eratosthenes as a significant topographical point in his structure of the world (F37, 48, 51–2, 55–6, 60, 62–4, 77–80, 83–6, 108). For the "sealstones" see F66.

F49. *Summary*: **The inhabited world is divided into two parts by a line from the Tauros to the Pillars. India is well defined geographically and ethnically, and is four-sided and rhomboidal.**

Commentary: How to separate and divide territories was a basic problem in geographical research, since political boundaries were not always meaningful (see F33), but India had the advantage of having distinct limits and ethnicity. The word "rhomboidal" (ῥομβοειδής) is from Euclidean geometry (*Elements, Definition* 22), a valuable tool for Eratosthenes.

F50. *Summary*: **Using many sources, but especially Patrokles, it can be determined that Meroë and southern India are on the same parallel, as well as the distance to the parallel of Athens, the width of the mountains and that it is the same distance from Kilikia to Amisos, and that it is a straight line east from Amisos through Kolchis to the eastern sea, and straight west to the Hellespont.**

Commentary: Hipparchos (his F12) complained that Eratosthenes relied solely on Patrokles (see F47) for the dimensions of India, an odd charge since his use of Megasthenes, Deimachos, and others is apparent (F67, 69, 73, 75). Even Strabo found this peculiar, listing a number of items about India not from Patrokles (although unfortunately the actual sources are not cited). For the reasons for Hipparchos' charge, see Dicks, *Hipparchus* 123.

F51. *Summary*: **From Amisos to Kolchis is due east, as is the route to the Kaspian and Baktra, a fact demonstrated by crops and**

winds, although ancient maps have it at a slant: in fact there are many errors in these maps.

Commentary: Although it is not clear whether the statement is from Eratosthenes or Strabo, the reference to Euclidean geometry (see also F49) regarding parallel lines (quoting Euclid, *Elements*, *Definition* 23) demonstrates Eratosthenes' dependence on it for establishing his grid, although the argument is somewhat of a syllogism, since convergent lines may appear to be parallel on the vastness of the earth's surface. Eratosthenes has listed a number of criteria for determining that two points lie on the same latitude, largely information obtained from sailors and (beyond Kolchis) those using the trade routes. Eratosthenes' eastern point, Baktra, the eponymous capital of Baktria, had been visited by Alexander and was still an important city (Frank L. Holt, *Thundering Zeus: The Making of Hellenistic Bactria* [Berkeley 1999] 124–5, 128–9). The ancient sailors' wisdom that crops and winds were determinators of latitude was useful in the Mediterranean and Black Sea, but was hardly applicable to the lands beyond, where solar position was the most reliable criterion. Most of the passage is from Hipparchos or later: the range of mountains running across Europe from the Pyrenees to the Balkans had long been vaguely known, if strangely. Herodotos believed that Pyrene was a village near the source of the Danube (2.33) and Alpis a central European river (4.49). It is only with Aristotle that the Pyrenees are located as a mountain range in western Europe (*Meteorologika* 1.13): Eratosthenes may have been the first to realize that the Alps were mountains. The major anachronism in the passage is the German mountains, a term not even used by Hipparchos but first documented by Poseidonios (his F73 Kidd [=Athenaios 4.153e]) yet still highly uncertain (I. G. Kidd, *Posidonius* 2: *The Commentary* [Cambridge 1988] 323–6). Even Tacitus, over a century later, found the term "Germania" new and recent (*Germania* 2). Eratosthenes would have used "Keltika" to apply to this region. The fragment concludes with a comment about how the ancient plans (again unspecified, as in F47) were full of mistakes.

F52. *Summary*: **The route from Thapsakos to the Kaspian Gates is 10,000 stadia in a straight line. The Kanobic mouth and the Kyaneai lie on the same meridian, 6,300 stadia from the Thapsakos meridian. Mt. Kaspios is 6,600 stadia from the Kyaneai, so Thapsakos and Mt. Kaspios are on essentially the same meridian. All these distances are loosely calculated.**

Commentary: Hipparchos (his F29–31) objected to one of Eratosthenes' basic distances, that from Thapsakos to the Kaspian Gates, although Strabo came to his defense in noting that his measurements were loose, not mathematical (which, of course, was Hipparchos' basic objection). The argument is clear from the text except it is not stated that Eratosthenes had made it only 7,400 stadia from the Gates to Mt. Kaspios (this is in F108). Since Mt. Kaspios is equidistant from both the Gates and Thapsakos, the figure of 10,000 stadia had to be circuitous. Thapsakos ("The Crossing") is the ancient crossing of the Euphrates, used by Alexander, but abandoned thereafter, surviving only as a major point in Eratosthenes' scheme, and not certainly located today (see Michal Gawlikowski, "Thapsacus and Zeugma: The Crossing of the Euphrates in Antiquity," *Iraq* 58 [1996] 123–33; Ronald Syme, *Anatolica: Studies in Strabo* [ed. Anthony Birley, Oxford 1995] 97–9). Mt. Kaspios is a general term for the Caucasus mountains. For the Kaspian Gates see F48. Hipparchos found further fault with Eratosthenes' topographical data at the eastern end of the Black Sea, and may have been aware that he had misplaced Dioskourias, which is not at the eastern end of the sea. Dioskourias and Phasis were Milesian settlements and had become large and important cities by Hellenistic times, but Dioskourias was northwest of Phasis, which itself was at the mouth of the river of the same name (the modern Rioni) and was the point of access to the interior and the routes to the Caspian Sea, which could not be gained from Dioskourias, as Eratosthenes would have it. The itinerary Bosporos-Phasis-Dioskourias may reflect a shipping route. See David Braund, *Georgia in Antiquity* (Oxford 1994) 30–3, 96–109, and Dicks, *Hipparchus* 43–4.

F53. *Summary*: **The inhabited earth is in the shape of a chlamys, tapering at its extremities. Its north-south width is delineated from the Kinnamomophoroi to the parallel through Ierne, and the east-west length from the Pillars to the Eastern Ocean beyond India. Taprobane is well south of India, and the region around Ierne is far to the north. The western boundary is the Sacred Promontory, in Iberia.**

Commentary: For the chlamys shape, see F34. Eratosthenes laid out the dimensions of the inhabited world, with his main meridian from the Land of the Kinnamomon Bearers (or the preferable Island of the Egyptian Fugitives, see F35) north to Ierne, which is mentioned only here and in F37. Ierne (Ireland) was probably not known to Eratosthenes (the earliest citation is Caesar, *Gallic War* 5.13 [as Hibernia]; but see

M. Cary and E. H. Warmington, *The Ancient Explorers* [revised edition, Baltimore 1963] 49), although vague previous references to the "Sacred Island" ("Hieronnesos") may be a confusion of the toponym (see Philip Freeman, *Ireland and the Classical World* [Austin 2001] 28–35), and its use may be Strabo's convenient substitute for Pytheas' Thoule, the normal northern end of the main meridian (F35). The main parallel was outlined in F47. Eratosthenes stressed the validity of his chlamys theory, and provided a new equivalent latitude: Taprobane (see F74) and the Island of the Egyptian Fugitives. Although the location of the latter is not known with certainty (see F35), if around Khartoum it would be about 8° north of Taprobane, reinforcing the placement of India too far north (see F47). As support to a comparable latitude, Eratosthenes noted the traditional reasoning that the temperature was similar. The data about the Island presumably came from the Ptolemaic explorers of the upper Nile; Taprobane was known through Seleukid exploration. Eratosthenes also listed some comparisons at the northern extremity of the inhabited world: the "outlet" of the Hyrkanian (Caspian) Sea was north of farthest Skythia, and Ierne (perhaps again a substitute for Thoule) was still farther. For the Caspian Sea as an inlet of the Ocean see F24; its "outlet," presumably at the north end near the mouth of the Ural River, is on the latitude of Paris, so the basic scheme is accurate. *Oroskopeia* was a method of using a sundial that had been marked with curves for solstices and equinoxes and a gnomon in order to fix latitude (Geminos, ed. Evans and Berggren 134–5), a common but error-prone technique (Kidd, *Commentary* 732). Using winds to determine latitude (see also F51) was sailors' lore. The best method of establishing latitude was the length of the longest days and nights, which was 14½ hours along the main parallel (see F60). This allowed extension of the parallel beyond the Pillars to the ancient Carthaginian outpost of Gadeira (modern Cádiz) and the Sacred Cape (modern Cabo de San Vicente).

F54. *Summary*: **One can perceive differences in latitude as little as 400 stadia. Greater distances can be comprehended by differences in flora, fauna, and weather, and lesser ones by instruments.**

Commentary: Eratosthenes continued discussing the method for determining latitude (see F51, 53). For greater intervals one could rely on the traditional meteorological methods (less reliable outside the Mediterranean than Eratosthenes realized), but lesser ones had to be determined by instruments, which could distinguish the 400 stadia between the parallels of Athens and Rhodes (the figure is peculiar, as the distance is actually about 175 km.).

F55. *Summary*: **The inhabited world, measured loosely, is a parallelogram.**

Commentary: Strabo (probably following Hipparchos) expressed concern over Eratosthenes' lines from the Kaspian Gates to the "mountains" (presumably the Caucasus) and to Thapsakos (which, as noted in F52, are at an angle). Despite the unknown position of Thapsakos and the loose term "the mountains," the divergence is perhaps 500 km. Running a line from Thapsakos to Egypt (presumably Alexandria) is even more of a problem, as it would be due southwest. Thapsakos, an important point in Eratosthenes' system, caused many problems: see further F52 and Dicks, *Hipparchus* 130–6.

F56. *Summary*: **The length of the inhabited world is on a line from the Pillars to the Kaukasos. The third section is on the line from the Kaspian Gates to Thapsakos, and the fourth that from Thapsakos to the mouth of the Nile. From Rhodes to Alexandria is about 4,000 stadia.**

Commentary: In establishing his grid of parallels and meridians, Eratosthenes created a rigid sectioning of the inhabited world that did not conform to the realities of topography. This problem is especially apparent in his line from Thapsakos to Egypt (see also F55), which he considered necessary to separate Asia and Libya, following his sealstone concept (see F66). Although the exact location of Thapsakos is not known, its meridian would have been several hundred kilometers east of that of Egypt, so Eratosthenes was caught in the paradox of the impossibility of dividing the world by parallel lines. Hipparchos was quick to realize this (his F30) and it formed the basis of much of his dismissal of Eratosthenes' technique.

F57. *Summary*: **The Kinnamomophoroi (who once hunted elephants) are near the midpoint between the equator and the summer tropic. They are the southernmost to see the Little Bear in its entirety. To their east is the outlet of the Arabian Sea. Their parallel is slightly south of that of Taprobane and the same as southern Libya.**

Commentary: The data from the southern end of the main meridian are the same as in Book 2 except that here it is 3,000 stadia from Meroë to the Kinnamomon Bearers but 3,400 in F35. This adjustment is probably the work of Strabo, based on Hipparchos' astronomical observations (his F43). The Kinnamomon Bearers are described in slightly more detail than in F35, providing their actual location in the Horn of Africa and their reputation as elephant hunters, something of interest to the early Ptolemies. Placing their parallel slightly to the south of Taprobane increases the error of F53. On the other hand, the parallel is extended to the west across Libya but without any specific location (it would reach the Atlantic around Senegal): even though Greeks had been down the west African coast, no latitudes seem to have been taken.

F58. *Summary*: **From Syene to the equator is 16,800 stadia.**

Commentary: The figures reported are the same as in F35, but here the emphasis is on the terrestrial zones, placing the central burned zone at the southern end of the inhabited world (not, as one might expect, the tropic). This makes the burned zone 17,600 stadia across. There seems to be no concern here as to whether there is one or two burned zones, or an additional equatorial one (see F44–5).

F59. *Summary*: **The longest day at Meroë is 13 hours, which is halfway between the equator and Alexandria. At Syene the longest day is 13½ hours, and almost the entire Great Bear is visible. The parallel through Syene runs 5,000 stadia south of Kyrene, through Gedrosia and India.**

Commentary: Using the lengths of the longest day was an easy way to calculate latitude, and Eratosthenes seems to have collected data (mostly by report rather than personal effort) from as many places on his grid as possible. Turning these into actual latitude figures was probably something he could not do: this depended on the astronomical skill of Hipparchos (Dicks, *Hipparchus* 168–9, 192–3). For Ptolemais, Berenike, and Trogodytika, see F41. Because the ancient day was always divided into 12 hours and thus the length of hour varied, the standard became the equinoctial hour, $\frac{1}{12}$ of the length of a day at the equinoxes. Another way to determine latitude was the relative height of celestial bodies other than the sun, something that became apparent when Greeks began to venture regularly outside the relatively narrow latitudinal limits of the Mediterranean. Pytheas was aware of these changes in the far north, and took particular interest in the location of the celestial pole (Hipparchos, *Commentary on the Phenomena of Aratos and Eudoxos* 1.4.1; Strabo 2.5.8, 4.5.5; Kleomedes 1.4). In the south, the journey of Alexander took Greeks farther south than ever before (the mouth of the Indos is at 25° north latitude), and the disappearance of familiar constellations was recorded by Onesikratos (*FGrHist* #134, F10). In fact he probably supplied the comparative data about the eastern reaches of the parallel of Syene: the Fish-Eaters coast of Gedrosia (Onesikratos F28) lies about a degree north of the parallel of Syene. Since Kyrene and Alexandria are on essentially the same latitude (F60), it is not a notable feat to remark that the Syene parallel is 5,000 stadia south of Kyrene, in lands totally unknown (the Kufra Oasis region of southeastern Libya), but it does demonstrate how Eratosthenes' system could be used to plot inaccessible places.

F60. *Summary*: **The longest day is 14 hours 400 stadia south of Alexandria and Kyrene (which are 1,300 stadia south of Karchedon). The parallel of Alexandria and Kyrene runs through interior Maurousia, and east to India. At Phoenician Ptolemais the longest day is 14¼ hours, and it is about 1,600 stadia north of Alexandria. In the Peloponnesos it is 14½ hours, on a parallel 3,640 stadia north of Alexandria. At Alexandria in the Troad, and between Rome and Naples, the longest day is 15 hours, on a parallel 7,000 stadia north of Alexandria and 1,500 south of Byzantion. To the north is a parallel through Lysimacheia. Around Byzantion it is 15¼ hours, and 1,400 stadia north of the entrance to the Pontos it is 15½ hours, halfway between the pole and equator.**

Commentary: Much of the material in this fragment is Hipparchos' corrections and refinements of Eratosthenes' raw data (Hipparchos F48–57). Eratosthenes probably could only supply the length of day and some of the stadia measurements, but the material is tangled with the recensions of both Hipparchos and Strabo. Eratosthenes probably placed the 14-hour (for the meaning of "hour" see supra, p. 169) day through Alexandria, and put Carthage too far south (since Hipparchos questioned its gnomon data), although the 1,300 stadia interval is not between Carthage and Alexandria but between Carthage and the line 400 stadia south of Alexandria, as is apparent below. Carthage is actually over 1,000 km. north of the latitude of Alexandria. As he often did, Eratosthenes extended the parallel as far west and east as he could: if he actually mentioned Maurousia (see also F39, 100, 107), it is the earliest citation of this vast region of northwest Africa, all the territory beyond the Carthaginian hinterland (modern Algeria and Morocco). In Eratosthenes' time it was barely known to the Greek world, ruled by an indigenous kingship allied with Carthage that was already causing difficulty for the Romans (Diodoros 13.80.3).

Eratosthenes' next parallel was through Ptolemais in Phoenicia (Tyre and Sidon are about a degree of latitude to the north), another foundation of Ptolemaios II (Getzel M. Cohen, *The Hellenistic Settlements in Syria, the Red Sea Basin, and North Africa* [Berkeley 2006] 213–21). The next one ran through the middle of Rhodes (with a longest day now 14½ hours). This was essentially his prime parallel but some alteration is apparent (presumably the work of Hipparchos): the 3,640 stadia north of Alexandria does not agree with Eratosthenes' own calculation of 3,750 (F128), and it passes across the southern part of Sicily, not the Straits of Messina.

The next parallel is focused on Alexandria Troas, the city that Antigonos I founded near Troy. The longest day here was 15 hours. Nikaia, mentioned nowhere else in the extant *Geographika* and the hometown of Hipparchos, may show that he adjusted this parallel, as Eratosthenes is not cited, although mention of Rome, if by Eratosthenes, is one of his few references to that city (see also F65, 155). Yet Italy is placed somewhat to the north of where it should be. It seems that Eratosthenes used the new Hellenistic cities as much as possible for his central points, because the next parallel is based on Lysimacheia, a foundation of Alexander's companion Lysimachos. There is no recorded attempt to extend this parallel to the west, as data were probably lacking, although it is astonishing that there was no point in Italy that could be used. It runs close to Naples, but since the previous parallel allegedly ran north of Naples, it would be assumed that this one ran north of Rome, where there were probably no places known to Eratosthenes. Beyond, it runs across Sardinia and central Iberia, close to the site of Madrid.

The next parallel was that of Byzantion (again the data were probably adjusted by Hipparchos [his F52]), and the final one is in an unspecified point in the Black Sea. A longest day of 15½ hours would be near Pantikapaion at the mouth of the Maiotic Lake, modern Kerch, a region that Eratosthenes was curious about (F61), but not previously mentioned as a parallel (in F34 the parallel is placed at the Borysthenes, about 2° north). Placing the parallel at Pantikapaion has the advantage of being halfway between the equator and the pole (the day of 15½ hours is about 20 km. north of this line), and Eratosthenes may have decided to include this additional parallel because of the data he had on the site. Yet the text has no obvious connection to him because Strabo was probably working directly from Hipparchos (his F56). Eratosthenes' northernmost parallel, that through Thoule, was ignored by Strabo because of its connection to Pytheas. On Eratosthenes' use of the length of day, see Aubrey Diller, "Geographical Latitudes in Eratosthenes, Hipparchus and Poseidonius," *Klio* 27 (1934) 254–9.

F61. *Summary*: **At Pantikapaion, in the Temple of Asklepios, is an epigram remarking on the severe winters.**

Commentary: Pantikapaion was a Milesian foundation at the mouth of the Maiotic Lake on the north coast of the Black Sea, at modern Kerch (see also F60). In Eratosthenes' day it was an outpost of the Bosporan kingdom. It also lay halfway between the pole and the equator. It is doubtful that Eratosthenes had been there, since he does not seem to have traveled much for research purposes, working, as Hipparchos noted (F50) in the world's finest library, which no doubt had collections of epigrams. There is no hint of the date of the epigram. The weather at Pantikapaion would have seemed strange to those from the Mediterranean: it was one of the three northernmost Greek cities (only Tanais and Borysthenes were farther north), and the region was noted for its severe winters: Mithradates the Great reported ice on the sea in summer, and it was beyond the limit of the grapevine (Strabo 2.1.16), the ultimate indicator for a Greek of a place too far north.

F62. *Summary*: **The line from Thapsakos to Egypt is the extent of this portion of the inhabited world, and it is 6,000 stadia from Pelousion to Thapsakos, and 4,800 along the Euphrates from Babylon to Thapsakos.**

Commentary: The emphasis is on Hipparchos' objections to certain details of Eratosthenes, with renewed consideration of the problems with the line from Thapsakos to Egypt (see F55). Eratosthenes presumed that Thapsakos (for which, see F52) and Egypt were on the same parallel (in fact the parallels are several hundred kilometers apart, a major flaw in his system). He also had an overland distance from Thapsakos to Babylon, carefully noting that it followed the river. For the problems inherent in this fragment see Dicks, *Hipparchus* 130–5 and F63.

F63. *Summary*: **From Thapsakos to the Armenian mountains is over 1,100 stadia, and the mountains are on the parallel through Athens. The route from Thapsakos to Babylon is along the Euphrates: the Euphrates and Tigris make a great circle that encloses Mesopotamia. Distances in the Armenian mountains are either unmeasured or only approximated.**

Commentary: As with F62, Hipparchos continued to belabor issues about Eratosthenes' distances in Mesopotamia, and there is only a little of Eratosthenes' data buried within this text. It greatly disturbed Hipparchos that Eratosthenes measured from Thapsakos to Babylon along the Euphrates (which makes an arc to the west) rather than casting a straight line. Even Strabo found this objection somewhat pedantic and noted that in the mountains near Armenia one could only assume distances, as no measurements existed. Part of Eratosthenes' difficulty was the breakdown of the rigid mathematical figures he had been able to create for regions farther east (e.g. the rhomboid shape of India, see F49, 64), and the fact that as his researches moved into Mesopotamia there were far more data available along existing roads and trade routes, which he incorporated into his structure. Hipparchos' complaint that the latitude of Athens did not conform to that of the Tauros and Armenia is valid to a point (the western end of the Tauros, near Tarsos, is at the latitude of Athens but the eastern parts and Armenia are well to the north). As in F48, Eratosthenes placed Armenia too far south. See also F52.

F64. *Summary*: **The Indos flows south (although old maps have it running southeast), parallel to the meridian through the Kaspian Gates. India is rhomboidal in shape, with its eastern side pulled to the east.**

Commentary: Eratosthenes' prime meridian was outlined in F35. Other meridians (unlike the parallels) were more loosely constructed, since the shorter north-south extent of the inhabited world meant there were fewer data points than for the parallels, and moreover Strabo's recension may have deemphasized the meridians. East of the prime meridian of Alexandria and Rhodes there was a significant one through the Kaspian Gates (for which see F48): unfortunately there were no fixed points to anchor it either to the north or south. Nothing is said about what might be north of the Kaspian Gates (believed to be the end of the civilized world): to the south there is only a vague reference to the boundary of Karmania and Persis, presumably somewhere in western modern Kerman (Hipparchos [his F24] demonstrated that this could not be an actual meridian, as the line runs southeast). Eratosthenes also seems to have made the Indos River into a meridian, assuming it ran due south. Hipparchos corrected this to southeast (perhaps anxious to uphold the "ancient maps" that Eratosthenes usually mistrusted [see F51]), but the portions of the river known to contemporary Greeks flow south-southwest (actually from its source near Gartok in Tibet it flows northwest for over 400 km.—this portion was probably unknown to Greeks—and then turns sharply to the southwest across Pakistan). For the rhomboidal shape of India see F49.

F65. *Summary*: **It is 900 stadia from Epidamnos to the Thermaic Gulf, and 13,000 from Alexandria to Karchedon. Rome is on the same meridian as Karchedon.**

Commentary: The distances listed are all ones that Strabo found to be incorrect, and all are indicative (to Strabo) of Eratosthenes' ignorance of western regions. The distance from Epidamnos to the Thermaic Gulf (probably Thessalonike), using the ancient Corinthian trade route that had existed for centuries, is about 300 km. and thus can hardly be as small as 900 stadia. A century after Eratosthenes it became the Via Egnatia, one of the major Roman routes to the east, and so was much better known in Strabo's day. Strabo also pointed out conflicting data concerning Eratosthenes' western meridian: sailing from Karia to Porthmos (the Strait of Messina) should be the same distance as from Alexandria to Carthage, as the same interval is involved. An error of 4,000 stadia over 13,000 was unacceptable, Strabo felt. Yet the discussion reveals the peculiar western meridian of Eratosthenes: from Carthage to Porthmos to Rome, which, as Strabo pointed out, showed deep ignorance of the region. What Strabo did not say was that the distances are proportionally correct because Porthmos is well east of Carthage (about 450 km.: in fact, one sails due east from Carthage to reach the southern tip of Sicily), so Eratosthenes fell into the trap of taking existing (and largely correct) sailing distances and forcing them into a nonexistent meridian. He seems to have been unaware of the slant of Italy, assuming that the peninsula ran due north-south, and was north of Carthage. In fact Carthage is on the longitude of Pisa, so a line Carthage–Porthmos–Rome would be an immense arc to the east. Interestingly Vergil at the opening of the *Aeneid* (1.13–14) reflected Eratosthenes' view in placing Carthage and the mouth of the Tiber opposite one another, demonstrating Vergil's wide scholarly reading: see Martin Korenjak, "*Italiam contra Tiberinaque longe / Ostia*: Virgil's Carthago and Eratosthenian Geography," *CQ* 54 (2004) 646–9.

F66. *Summary*: **The main parallel divides the inhabited world into two portions. These can be further divided into "sealstones," with the first India and the second Ariana.**

Commentary: Having established his basic grid, whatever its known and unknown faults, Eratosthenes then divided the inhabited world into units that he called "sealstones" or "gemstones" (σφραγῖδες), a perfectly common Classical Greek word. He seems to have intended it to describe an irregular quadrilateral, applying best to the perceived shape of India (F49): "sealstone" may have been a word more in the vernacular than the Euclidean technical term of "rhomboid." The sealstone concept began to degenerate after the first two (Ariana and India) simply because topographical and ethnic regions could not be defined by geometry, however loosely applied. Eratosthenes' use of τινα and τι near the end of the fragment indicate his uncertainty even for India.

F67. *Summary*: **Deimachos, in contrast to Megasthenes, is unreliable about India, as he says that the Bears are not hidden and shadows do not fall to the south, something that happens only 5,000 stadia south of Alexandria.**

Commentary: The first sealstone, and where the concept worked best, was India. Eratosthenes relied on several existing reports, especially Deimachos (*FGrHist* #716) and Megasthenes (*FGrHist* #715), both early Hellenistic envoys to the Mauryan court at Pataliputra (see F22). Two latitudes and locations of India are provided: that of Deimachos is more accurate since India does not extend into the southern hemisphere, but Eratosthenes was correct in confirming Megasthenes' view that the Bears are hidden in southern India (they begin to disappear at the tropic and set by Cape Comorin, the southern end [Dicks, *Hipparchus* 127, 172–4]). The passage is typical of Strabo's complex negative arguments about his sources. On this passage see Strabo, *Géographie* 2 (ed. François Lasserre, Paris 2003) 132; on Eratosthenes and India, see Klaus Karttunen, *India and the Hellenistic World* (Helsinki 1997) 100–5.

F68. *Summary*: **Both Bears set in India, as Nearchos showed. Deimachos did not believe this, but Megasthenes did.**

Commentary: Philon (see F40) was probably the first to suggest a comparative latitude between the upper Nile and India, using data from Nearchos (*FGrHist* #133), the commander of the fleet that sailed from the Indos to rejoin Alexander at Sousa. Yet Philon was in error, because Meroë is over 8° north of the southern tip of India, and this created an India that is truncated and too far north. Obviously Eratosthenes preferred Deimachos' latitudes (see F67), but Strabo's polemic makes it difficult to separate out the five sources mentioned in this short fragment.

F69. *Summary*: **The Indos was the western boundary of India at the time of Alexander. Ariana was under Persian control at that time, but later was taken by the Indians. The boundaries of India are the northern mountains, the Indos, and the Ocean, creating a rhomboidal shape. The east-west extent is known as far as Palibothra, but less so beyond, as noted in a record of stopping points, although the entire distance is at least 16,000 stadia.**

Commentary: Eratosthenes' summary of the shape of India is almost entirely from Megasthenes (*FGrHist* #715), whose report was based on data collected at the court of Chandragupta at Pataliputra (Greek Palibothra or Palimbothra), between 318 and 305 BC (it was not until the latter year that Chandragupta gained territory west of the Indos, described as an event after Megasthenes' presence: see F71). The description begins by following the mountain range from the Tauros (in Anatolia and thus in familiar territory) through the Parapamisos (the Hindu Kush) and to the Emodos and Imaos (variant terms from two sources and preserved as the Himalayas: see Klaus Karttunen, *India and the Hellenistic World* [Helsinki 1997] 107–8), thereby placing India in a relationship to the known world. For why the Makedonians called the range the Kaukasos, see F23–4. The concept of a great east-west mountain range dividing the inhabited world in two may be the creation of Eratosthenes, although there are hints of it in Dikaiarchos' geographical scheme (his F123 [= Agathemeros, *Prologue* 5]; see Karttunen 106). The suggestion that India borders the Atlantic on the east connects with the contemporary belief that one could sail west from the Pillars to India (F33: see Aristotle, *Meteorologika* 2.5), although the idea that the Indos is the western boundary of India probably goes back to Achaimenid times. (Klaus Karttunen, *India in Early Greek Literature* [Helsinki 1989] 36).

Eratosthenes affirmed Megasthenes' calculation of 16,000 stadia for the north side, and 13,000 for the west side. The figure of 16,000 may be what Pliny converted to 1,875 miles (F70), which would be 2,775 km., surprisingly close to the actual distance from the mouth of the Indos to that of the Ganges. By the same calculation the west side would be 2,250 km., about the distance from the mouth of the Indos to Cape Comorin (although not, as stated, the length along the Indos itself, which is far less).

The account then focuses on Pataliputra, indicating that Megasthenes is the probable source, and noting that the distance "as far as" (perhaps from the Indos to) Pataliputra had been measured and there was a royal road for 10,000 stadia. Little was known about the Ganges below Pataliputra. The figure of 16,000 stadia is specifically attributed to Megasthenes and "a record of stopping points," of unknown date, presumably an itinerary along the road, perhaps from Baiton (*FGrHist* #119), who did the route measuring for Alexander and who seems to have been in India at a later time (see Pliny, *Natural History* 6.69). For Patrokles, see F47. The Koniakoi (mentioned only here and in F74) are specifically placed at the southern tip of India, one of several details about the south learned by Megasthenes through hearsay. On this passage see Nicola Biffi, *L'estremo oriente di Strabone* (Bari 2005) 154–5.

F70. *Summary*: **The coastline of India, from the east to the Indos, is 4,350 miles.**

Commentary: Pliny's recension of Eratosthenes' stadia into Roman miles provides a vague control on Eratosthenes' distances (see F69), using the standard equivalent of 1,480 m. to a Roman mile, providing about 6,450 km. for the entire coast of India.

F71. *Summary*: **Megasthenes, who often visited India, considered it the largest portion of Asia. The region between the Euphrates and the Indos is much smaller. India is bounded by the Ocean, the mountains, and the Indos, and is mostly an alluvial plain.**

Commentary: Arrian's brief summary of Eratosthenes' description of the position of India covers the material in F69, with a vague reference to the four sealstones of southern Asia. The passage is significant because it provides the fullest extant biographical datum about Megasthenes (*FGrHist* (#715), almost certainly from Eratosthenes, connecting him with Sibyrtios, the virtually independent satrap of Arachosia between 325 and 316 BC, and Chandragupta (Greek Sandrakottos), who ruled 318–294 BC at Pataliputra (Greek Palimbothra or Palibothra). This demonstrates that on occasion Eratosthenes included biographical information about his sources. Otherwise the account is generalized

and adds nothing new, even taking for granted the Makedonian mis-
placement of the Kaukasos (F23–4). See further, Bosworth, *Commentary* vol. 2, pp. 236–46.

F72. *Summary*: **The west side of India is 13,000 stadia, and the opposite side is 16,000. There is a royal road across India as far as
Palibothra: less is known beyond there.**

Commentary: In his *Indika*, Arrian summarized more fully than in
his *Anabasis* Eratosthenes' data about India (see F71). The measurements conform to those quoted by Strabo (F67, 69). The Indos River has
its source in Mt. Tauros (as expansively defined by Eratosthenes) and it
is noted that the royal road to Pataliputra was measured in *schoinoi* (for
which see F27), which would mean that it had been surveyed by someone familiar with Mediterranean units of distance, perhaps Alexander's
surveyor Baiton (see F69).

F73. *Summary*: **The reliability of Patrokles is questionable, because he disagrees with Megasthenes about Indian distances,
but neither is as reliable as the record of stopping points.**

Commentary: For Patrokles and the problems with his reliability, see
F47. Eratosthenes had access to a "published record of stopping points,"
although he did not specify the author or exact source. Such itineraries
became common in the Hellenistic period (an overland counterpart to
the coastal sailing narratives or *periploi*), and several are extant: the
best example is the *Parthian Stopping Points* of Isidoros of Charax
(*FGrHist* #781), which preserves an itinerary in *schoinoi* (see F27, F72)
from Zeugma on the Euphrates to Alexandria in Arachosia, written in
the Augustan period but incorporating earlier data, probably from some
of the same sources used by Eratosthenes.

F74. *Summary*: **All of India is watered by rivers, of which the two
largest are the Indos and Ganges. India is flooded by summer
rains and has two harvests. The fauna and people are similar to
those of Aithiopia and Egypt. Taprobane is seven days south of
India.**

Commentary: Eratosthenes' longest extant description of India is perhaps a synthesis of the four sources mentioned by Strabo (15.1.12–16)
before and after this fragment: Onesikratos (*FGrHist* #134), Nearchos

(*FGrHist* #133), Megasthenes (*FGrHist* #715), and Deimachos (*FGrHist* #716), who were quoted by Eratosthenes (see F22, 68). Megathenes was probably used the most. There has been some tampering with Eratosthenes' text, either by Strabo or a predecessor, as he would not have called the mountains north of India the Kaukasos, since he had already emphasized this nomenclature was erroneous (F23–4).

Greeks were overwhelmed by the many characteristics of India that they found alien, although to some extent there was a tendency to speak of India in the ideal terms they often used about foreign societies. But Greeks could not fail to be impressed by the striking differences that contrasted India with their arid homeland, especially the great rivers and agricultural wealth. The Indos had been known since the sixth century BC (Herodotos 4.44) but Greek travelers after Alexander had come across an even larger river (so it seemed), the Ganges. Not even Alexander had seen it, although he received vague reports, and in later years knowledge of the Ganges became tangled with the idea that Alexander had reached the Eastern Ocean, yet Megasthenes was probably the first Greek to reach the river (Bosworth, *Commentary* vol. 2, pp. 339–41, 347). That India could be even more fantastic than Alexander realized only added to its mystique. Actually the Ganges is not as long as the Indos (2,500 vs. 2,880 km.), but much of the upper course of the latter is deep in Tibet and was never seen by Greeks, and the Ganges has the appearance of being larger because of its volume. Because the lower Ganges was only known by report, its massive multiple mouths, impressive to the modern traveler, were totally unknown to Greeks. The mouths of the Indos were better known as they had been seen by Alexander and his entourage: Eratosthenes' source here is probably Nearchos, who had to navigate them. The number varies in different accounts, with as many as seven reported in the first century AC (*Periplous of the Erythraian Sea* 38), but the two cited by Eratosthenes were certainly the main ones. It was perfectly reasonable to compare the Indos delta to the other great river delta known to Greeks, the Nile. The issue of which delta was larger appears at least as early as Strabo (15.1.33), who objected to Onesikratos' statement that they were of equal size (his F26): what Eratosthenes believed is not clear. He may also have learned (from Nearchos or Onesikratos) about the notable city of Patala on an island in the delta (Strabo 15.1.13), which gave its name to the district of Patalene.

The agricultural productivity of India was also a matter of great curiosity to Greeks, especially the role of the monsoons, when the entire countryside became a lake. Because of the large rivers, it was not difficult to assume that evaporation and the prevailing westerly winds brought the summer rains. Aristotle had already made it clear that evaporation and condensation were the cause of rain (*Meteorologika* 1.9). The two annual harvests were another alien phenomenon, as well as the wide variety of crops (*bosmoron* is a type of millet, mentioned only here and at Strabo 15.1.18 and [as *bosporos*] Diodoros 2.36, all three citations probably from Megasthenes [see his F4 and Andrew Dalby, *Food in the Ancient World from A to Z* (London 2003) 219]). India was a land of mythic abundance where even farmers were sacred (Megasthenes F19).

There were also unusual fauna, although these seemed similar to what Greeks had encountered in Egypt. The hippopotamus had been known to Greeks ever since Euthymenes of Massalia saw it in West Africa around 500 BC (Aristeides 36.85–95; see also Hanno's almost contemporary report), but has not existed in India since prehistoric times, despite Onesikratos' comment. There was a tendency to compare India to Egypt, as it was the only other tropical territory well known to the Greeks, seeing similarities in its rivers, flora, fauna, and people, but the analogy could be carried only so far as Indian phenomena went far beyond anything produced in Egypt.

Taprobane (Sri Lanka) first came into Greek knowledge at the time of Alexander. Onesikratos (see his F13) was probably the first to report on it, with Megasthenes (his F26) slightly later. There is no evidence that either visited the island: direct contact probably did not exist until the Roman period, and even the author of the *Periplous of the Erythraian Sea* (61), while acquainted with its commerce, does not seem to have known the place. The most lengthy description is that of Pliny (see F76), but much of it is from the ambassadors who came to Rome in the time of the emperor Claudius. Megasthenes had more details about the island than Eratosthenes seems to have transmitted. Taprobane (one of several names for the island known to Greeks and Romans) is perhaps Sanskrit "Tambraparni" (encountered by Greeks through Tamil "Tampirapanni"): see Lionel Casson, *The Periplus Maris Erythraei* (Princeton 1989) 230–2. The seven days' sail could hardly be directly across the narrow Palk Strait. The Koniakoi (see also F69) are little known.

F75. *Summary*: **Megasthenes noted the two yearly harvests in India and the annual rains. There are many unusual plants.**

Commentary: The peculiarities of Indian climate were of interest to the Greeks. Eratosthenes, relying on Megasthenes (his F8), emphasized the two annual harvests, the monsoons, unusual flora and fauna, and the natural process of ripening in the tropics. All these phenomena were alien to the Mediterranean. Also a curiosity to the Greek world was cotton, first mentioned by Herodotos (3.47, 106, 7.65): by early Hellenistic times it was extensively farmed in the Persian Gulf regions (Theophrastos, *Research on Plants* 4.7.7). It is normally described in Greek literature as "wool from trees," and the actual Greek word, probably κάρπασος, is rare and not documented before Roman times (e. g. Strabo 15.1.71).

F76. *Summary*: Taprobane is 7,000 by 5,000 stadia, and has no city, only villages.

Commentary: The only part of Pliny's description of Taprobane (see F73) specifically attributed to Eratosthenes is its dimensions and the lack of urbanization, another curiosity to the Greeks.

F77. *Summary*: Ariana, subject to Persia, is bounded, like India, by the mountains, the Ocean, and the Indos, extending as far west as the line from the Kaspian Gates to Karmania, and is shaped like a quadrilateral. There are several ethnic groups living within it.

Commentary: The second of Eratosthenes' sealstones (see F66) is Ariana, the vast territory between Mesopotamia and India. The ethnym Arioi was used in Media in the sixth century BC and later applied to a Persian satrapy (Herodotos 7.62, 3.93) located in the vicinity of modern Herat in Afghanistan, where Alexander founded Alexandria Among the Areioi. Yet to use this local toponym and ethnym to describe the entire region from India to Mesopotamia seems to have been Eratosthenes' idea, which did not take hold. Strabo was inconsistent in his use of it, with "Ariana" applied only when describing Eratosthenes' scheme (2.1.22, 31, 15.1.10) and "Aria" for the district around Herat (2.5.32, 11.8.1), although with some confusion in both cases (see F78). But to make his sealstone theory work, Eratosthenes needed a general term for the entire region from India to Mesopotamia (see Gherardo Gnoli, "ARIANH: Postilla ad *airyō.sayana*," *RSO* [1966] 329–34).

Whatever the onomastic issues, Ariana was easily defined by the northern mountains, the Indos, the ocean, and the eastern edge of Mesopotamia, with this last side the most uncertain. This was a large region, twice the size of India (although Eratosthenes did not realize this) and, like India, known only sparingly. Alexander had gone east along its northern edge, although he actually was outside Eratosthenes' Ariana after he went through the Kaspian Gates, remaining so for some distance, and returned along its southern coast. The western interior was known through Median and Persian history but most of the central portion remained unvisited by Greeks. Thus the Greeks were in the position of knowing only the perimeter, and it is not surprising that most of Eratosthenes' data concerns boundaries, routes, and some ethnic groups. The most detail in this fragment is the description of the coast from the mouth of the Indos at Patala (see F74) to the Persian Gulf. This is based on the account of Nearchos (*FGrHist* #133) who, in late 325 BC, commanded the fleet that made that cruise. Many fragments of his report are extant, quoted by Strabo, Pliny, and Arrian. Eratosthenes seems to have been primarily interested in distances, so much of Nearchos' ethnography was eliminated. Only a few ethnyms are cited: the Arbioi and their eponym, the Arbis River (probably the modern Hab in western Pakistan), the Oreitai to the west, and the Ichthyophagoi ("Fish Eaters," a common Greek term for coastal peoples). The sum distance, 12,900 stadia, does not agree with the parts (which total 13,900 stadia): some editors have adjusted the manuscripts, and others eliminate the first 1,000 stadia from the total as being "part of India," not the voyage. On the passage see Nicola Biffi, *L'estremo oriente di Strabone* (Bari 2005) 244–6.

F78. *Summary*: **Ariana is large, bordered by the Indos, the Ocean, the Parapamisos, and the western boundaries of Parthyene and Karmania. Its width is 12,000 or 13,000 stadia and its length is recorded in the treatise *Asiatic Stopping Points*, which outlines two routes. Parts of Persis and Media are included in Ariana. There are various ethnic groups within the territory. Seleukos gave some of the eastern portions to the Indians. Alexander passed through the northern part.**

Commentary: Eratosthenes' most thorough extant description of Ariana relies heavily on previous itineraries and catalogues of ethnic groups. The only source mentioned is a work titled *Asiatic Stopping Points*, presumably one of the several available itineraries across Asia (see F73).

The boundaries of Ariana were delineated in greater detail than in F77, with particular attention to the uncertain western border with Mesopotamia. The Indos River defined the north-south extent of the territory. The length of 12,000 or 13,000 stadia is close to the actual extent of the river, but presumes a river running essentially straight in a southerly direction (this would place its source somewhere around

Alma-Ata). Nevertheless it was known that the source of the river was somewhere beyond the mountains, as recorded by Arrian, *Anabasis* 5.4.1, where ἐπὶ τάδε τοῦ ὄρους means "on this [i.e. Greek, not Indian] side of the mountains"; see further, Bosworth, *Commentary* vol. 2, p. 223. Greeks would not have been aware of the northwesterly direction of the river's upper half, but may have had a figure for its entire length. Eratosthenes then used the *Asiatic Stopping Points* to provide an east-west length of Ariana (only 15,300 stadia, hardly more than its presumed north-south width), creating an almost square territory. There is a digression on the route of the itinerary, which, interestingly, uses Aria in its localized form, not the broader version favored by Eratosthenes. East of the Kaspian Gates there were two ways to reach India: more or less straight toward Baktra or curving south into Drangiane and Arachosia. The former and more direct way, roughly through southern Turkmenistan to modern Merv, on to Baktra (modern Balkh in Afghanistan) and southeast to the vicinity of Kabul, was probably the ancient trade route across this territory. The longer way, which swung to the south along the modern Iran-Afghanistan border into Drangiane (the modern Helmand basin) and then back northwest through Arachosia to join the other route near Kabul, had the authority that it was the way Alexander had gone, but was probably a conflation of local and regional routes. Significantly, however, the distance is provided for this one alone, probably the work of Alexander's surveyor Baiton (*FGrHist* #119). Eratosthenes also found it necessary to emphasize further his definition of Ariana, extending from Persis to Baktria and Sogdiana, noting that the entire region was similar linguistically, as is the case today.

Perhaps from another source is a catalogue of ethnic groups, first those along the Indos and then the ones farther west. It is more thorough in the east: information on the western portion of Eratosthenes' Ariana is almost totally lacking (perhaps because, as the heart of the Persian Empire, this was well known). As presented by Strabo the catalogue dates from the reign of Seleukos I (died 281 BC), after the arrangement he made with Chandragupta to cede the eastern Seleukid territories in return for a marriage alliance and a gift of elephants (Appian, *Syrian War* 55). This was around 305 BC: the elephants appeared in the battle of Ipsos a few years later (Plutarch, *Demetrios* 28–9), marking the introduction of the animal into Hellenistic warfare. Eratosthenes' source is too late to be Megasthenes: Deimachos is a possibility, or an unknown ethnography of the early eastern Seleukid empire, written sometime between 305 BC and the establishment of the Greco-Baktrian kingdom half a century or more later. It is a source

favorable to Alexander, since it uncritically included the movement of
the Caucasus (see F23–4). For Alexandria of the Areioi, whose exact location remains uncertain, see P. M. Fraser, *Cities of Alexander the Great*
(Oxford 1996) 109–13. On the passage generally see Nicola Biffi,
L'estremo oriente di Strabone [Bari 2005] 256–60; on Seleukos' arrangement with Chandragupta, see Klaus Karttunen, *India and the Hellenistic World* (Helsinki 1997) 260–4.

F79. *Summary*: **The western side of Ariana is mixed ethnically
and hard to delineate, although marked by a line from the Kaspian Gates to the promontories of Karmania.**

Commentary: Eratosthenes was aware of difficulties with the western
side of Ariana, since any delineation that was ethnically based was impossible. The issue was probably the territory around Sousa, the ancient
Elamite capital that had become a royal city of the Persians in the sixth
century BC. It lay at the eastern edge of the Mesopotamian plain, topographically part of it but culturally belonging to what Eratosthenes had
defined as Ariana. Eratosthenes' line from the Kaspian Gates to the
Straits of Hormuz (the mouth of the Persian Gulf), defined here a little
more precisely than in F64 (perhaps Strabo's gloss), is used as the west
edge of Ariana, but this excludes the historic Median and Persian territories and leaves a vast triangle neither in Ariana or Mesopotamia.
Strabo (and probably Hipparchos) certainly realized the problem (the
arguments continue in F80) but offered no solution. The degeneration of
the sealstone concept demonstrates that Eratosthenes did not lay out
his divisions of the inhabited world in advance, but worked from India
west with an idea in mind that he eventually abandoned when it proved
untenable. In addition, Eratosthenes was faced with the difficulties between reconciling the traditional ethnically based view of the world with
his new concept that made use of landforms and topographical units.

F80. *Summary*: **The line marked by the mountain range forms an
angle with the line from Thapsakos.**

Commentary: The negative thrust of this fragment is due to Strabo's refutation of Hipparchos' claims (his F21–2) about the confusion of Eratosthenes' boundaries, parallels, and meridians in this region. He did not
expect his sealstone boundaries to lie on meridians (whether determined
in advance or merely a north-south line): the northwest-southeast orien-
tation of Mesopotamia, and thus the western boundary of Ariana, makes

this impossible. The fragment also revisits some of the issues of the relationship between the Kaspian Gates, Thapsakos, and Babylon that had been discussed in F52 and 56.

F81. *Summary*: **The seacoast from the Indos ends at Karmania, which is substantially farther north, at the mouth of the Persian Gulf, with the coast of Arabia visible across the mouth.**

Commentary: In F79 Strabo had hinted that the Straits of Hormuz (the modern name reflects a survival of ancient Harmozai: see F94) should be the southwest corner of Ariana, something that does not appear in Eratosthenes' defining of its boundaries (F77). Eratosthenes was familiar with Harmozai and the mouth of the Persian Gulf, but whether he incorporated it into his perimeter of Ariana cannot be determined, as he is not named in this fragment. The primary source here is Nearchos (*FGrHist* #133), a continuation of the material in F77. The Straits of Hormuz lie only 2½° north of the mouth of the Indos, but it was on the Karmanian coast that Nearchos and his companions first began to see familiar constellations (Pliny, *Natural History* 6.98). For Arabia Eudaimon see F92.

F82. *Summary*: **The third sealstone cannot be easily defined, because of its uncertain boundary with Ariana and its irregular shape. The Euphrates flows through it.**

Commentary: Eratosthenes' third sealstone is Mesopotamia, but the concept was deteriorating even more than with Ariana, as the problem with the western boundary of Ariana (F80) also affected Mesopotamia. He used the Euphrates as the western boundary of the third sealstone (although Strabo believed, with some merit, that the Mediterranean coast would have been a better choice), creating a long narrow territory running northwest to southeast and lying only on the left bank of the Euphrates. This was unquestionably Mesopotamia, since there were relatively few settlements west of the river, and it by definition was only the territory between the two rivers, but this choice failed to create the rectilinear shape that, at least in the case of India and Ariana, seemed to be a requirement for a sealstone. Moreover, as Strabo pointed out later (2.1.31), the ancient Assyrian territory and Syria proper were excluded, although they would have created a more orthodox shape. There were also problems with the southern boundary, discussed in F83.

F83. *Summary*: **The third sealstone is rendered roughly, because of problems with the eastern side, the intrusion of the Persian Gulf, and the windings of the Euphrates. Mesopotamia is created by the proximity of the Tigris and Euphrates. All of this is represented loosely despite the collection of a large amount of data. The northern side is difficult to measure, and the southern side has the intrusion of the Persian Gulf. There are further difficulties with the western side. The Tigris and Euphrates flow from Armenia to the south and enclose Mesopotamia, and at one point they are only 200 stadia apart.**

Commentary: In addition to the issues concerning the eastern side of the third sealstone (F79), there was also the matter of its southern side, which was irregular because of the Persian Gulf. The intrusion of the gulf was a difficulty Eratosthenes seems to have been unable to resolve. There seems a certain ambivalence about how it related to the sealstone: he may have wanted to run a line from its mouth to Babylon, but he seems not to have realized that this would actually run up the eastern side of the gulf. Thus his line extended from Babylon to Sousa, Persepolis, and beyond, following a measured route that had existed from at least Persian times. This demonstrates that whenever possible he used road itineraries rather than lines or boundaries that went across uncharted territory, but it created a southern boundary that excluded almost all of ancient Babylonia (but see F84). Strabo pointed out that Eratosthenes avoided saying that this southern side was parallel to the northern, a subtle indication that he believed sealstones ought to be rectangular. He also knew that the Euphrates curved in a broad arc: continuing his use of inventive descriptive terminology, he said it was like a cushion on a rower's bench in a ship (ὑπηρέσιον: see Thoukydides 2.93, where the context reveals that it was the personal possession of the rower), presumably a crescent shape. Eratosthenes further acknowledged that he would have preferred to find a northern boundary to this sealstone but could not, because of a lack of measured routes. There may have been a file of itineraries and reports in the Alexandria library catalogued as a group without specific titles. Eratosthenes' reliance on these reports as well as the irregularities of shape all meant that the third sealstone was outlined very roughly, as Strabo noted.

Since the theoretical north side of the sealstone was in the uncharted mountains of eastern Anatolia, Eratosthenes instead used for his boundary much of Alexander's route from the Euphrates to the Kaspian Gates. In the summer of 331 BC Alexander had crossed the river at Thapsakos, and the Tigris shortly thereafter (Arrian, *Anabasis* 3.7). After engaging Dareios III at Gaugamela, he began his conquest of Mesopotamia and Persis, resuming his pursuit of Dareios the following spring in the vicinity of the ancient Median capital of Ekbatana, heading toward the Kaspian Gates (Arrian, *Anabasis* 3.19–21). Thus there were two segments available to Eratosthenes: from the Euphrates at Thapsakos to east of the Tigris near Arbela, and from around Ekbatana to the Gates. These were well recorded, and the intervening segment, from Arbela essentially to Thapsakos, would have been supplied from one of the many anonymous itineraries, as it was a long-existing route from eastern Assyria to Media. Yet Strabo pointed out that using this line as the northern edge of the sealstone left undetermined the territory to the north, Armenia and upper Media.

Strabo was uncertain how Eratosthenes defined the western side. It runs from Thapsakos to Babylon, which, if the southern side begins there, should be the southwestern corner of the sealstone, but Strabo had the line continuing on to Teredon at the mouth of the Euphrates. This city is impossible to locate precisely today because of vast changes in the topography of lower Mesopotamia: in Hellenistic times the coastline was perhaps as much as 200 km. further inland than today (John Hansman, "Charax and the Karkheh," *IrAnt* 7 [1967] 36–45), and even between the time of Eratosthenes and Strabo there may have been as much as 50 km. of silting. Teredon was probably somewhere in the vicinity of modern Basra and was an important seaport in Alexander's day, where the Tigris and Euphrates joined, and the starting point for much of the exploration of the Persian Gulf region, especially that by Androsthenes in 325–323 BC (*FGrHist* #711), a source with which Eratosthenes was familiar (F94). Making the southwest corner of the sealstone at Teredon is sensible, but a direct contradiction with the use of an existing road from Babylon to Persis as the southern boundary. The north end of the west side was undefined above Thapsakos, but here too Strabo was confused, because Thapsakos should be the northwest corner of the sealstone. Eratosthenes' knowledge of the upper Euphrates and Tigris was scant and erroneous (see F87, 89), especially within the remote and mountainous territory known as Gordyaiene, the area of the source of the Tigris, directly south of modern Lake Van in Turkey. Although technically part of the Seleukid empire, this region was not truly known to the Greeks and Romans until the campaigns of Lucullus in the first century BC, the probable origin of Strabo's description (16.1.24). The source of the Euphrates, north of Lake Van, was even less known, although it was vaguely understood that the Euphrates made a great western arc in its upper course (it first flows west for over 400 km. to the vicinity of modern Malatya, and then turns south and southeast). Eratosthenes also provided a few toponyms for the lower Euphrates, especially the "Wall of Semiramis," the name given to a number of ancient fortifications that ran between the Tigris and Euphrates where they came closest to each other (about 40 km. apart), just north of Babylon. These walls were known to Greeks from Xenophon's report in the *Anabasis* (2.4.12–13), who described a mud-brick and asphalt construction 20 feet wide and 100 feet high, lying four days from the great city of Opis, which had shrunk to a mere village less than two centuries later. Xenophon's fortification, which he called the "Median Wall," was one of many in the region, some of which were quite ancient. Nebuchadrezzar had built one, and, much earlier, Shu-Sin of the Third Dynasty of Ur (R. D. Barnett, "Xenophon and the Wall of Media," *JHS* 83 [1963] 1–26). It is quite probable that one of these walls would be associated with the semi-legendary Semiramis, who was known as a builder (Herodotos 1.184, 3.155).

F84. *Summary*: **The third section is bounded on the north by a line from the Kaspian Gates to the Euphrates, on the southern from Babylonia to Karmania, and on the western from Thapsakos to the outlet of the Persian Gulf. Some of these distances have not been measured.**

Commentary: In this fragment the southern boundary of Mesopotamia runs from Babylonia (rather than the Babylon of F83) to Karmania, which seems a middle ground between the two conflicting sealstone corners of F83, Babylon and Teredon. It may have been an attempt, probably before the time of Strabo, to resolve the contradiction, but since Babylonia is a territory several hundred kilometers across, it is much less precise. Mention of the "outlet" of the Euphrates suggests Teredon. Whether these three different versions of the southern boundary represent Eratosthenes' own editing within his text, or later attempts to reconcile difficulties, remains unknown. Strabo's own confusion is a contributing factor to the uncertainties.

F85. *Summary*: **The distance from Babylon to the Kaspian Gates is 6,700 stadia, and 9,000 to Karmania.**

Commentary: Needless to say, Hipparchos (his F23–4) made the most out of the problems with the boundaries of Eratosthenes' third sealstone, especially emphasizing that the distances reported created mathematically impossible figures (see Dicks, *Hipparchus* 131). He demonstrated that Eratosthenes placed the Kaspian Gates too far east, which would also be the case with the line south to Karmania. Since Eratosthenes relied on reported distances from a wide variety of sources, this is to be expected.

F86. *Summary*: **The length toward the north is about 8,000 stadia, and no more than an additional 2,000 to the Kaspian Gates. The width from Sousa to Persepolis is 4,200, and an additional 1,600 to Karmania. There are various ethnic groups in the territory.**

Commentary: The measurements in this fragment, although totaled somewhat differently, agree with those in F83–5. The passage also includes a brief and peculiar ethnography of Persian tribes. The Pateischoreis are not known elsewhere, although the name is Hellenized Persian in form. The Achaimenidai were one of the original Persian clans (Herodotos 1.125), but to speak of them as such would have been anachronistic in Hellenistic times, when the name had become dynastic rather than ethnic. The Magoi had long been familiar to Greeks as a quasi-religious and political group, but Herodotos (1.101) had recorded them as a Median clan. The Kyrtioi, whose name may be reflected in the modern Kurds, were a mountainous people noted for their semi-nomadic and predatory ways (Strabo 11.13.3), often mentioned in tandem with the Mardoi, whose nature was similar. As a whole the list is a strange mixture of the very old and very recent, perhaps summarized from an early list but with more recent additions.

F87. *Summary*: **Mesopotamia owes its name to its position, lying between the Tigris and the Euphrates, with the Tauros on the north. At their closest, the rivers are no more than 200 stadia apart. The Tigris flows through Lake Thopitis, which has some unusual phenomena.**

Commentary: The first part of this fragment is a summary definition of Mesopotamia. Strabo used the toponym several times, frequently citing Eratosthenes (see F83), and it is probable that Eratosthenes originated the term, since it seems to have been unknown to Herodotos and Xenophon, and is conspicuously absent from Berossos' history (*FGrHist* #680). The term may also have come from the era of Alexander. The earliest extant citation is Polybios (5.44.6). Much of the fragment is about the course of the Tigris from its source north of Lake Thopitis, the Thespitis of Pliny (*Natural History* 6.128), generally believed to be modern Lake Van, which is neither in the Tigris basin nor that of any river system except the small streams emptying into it. The source of the Tigris is actually south of the lake, although affluents are as close as 10 km. from its southeastern corner, so it is not difficult to see how it could be believed that the river ran underground from the lake for a few kilometers, especially in a rugged region whose topography was little understood. For the topography of the region, see T. A. Sinclair, *Eastern Turkey: An Architectural and Archaeological Survey* (London 1987) vol. 1, pp. 75–8. Rivers that allegedly ran underground were a common feature of Greek lore (see F140): the best example is the Nile, which, according to Juba II of Mauretania, originated in the mountains of his kingdom and flowed underground and through lakes to reach its known course south of Egypt. This theory was so persuasive that in the early nineteenth century explorers were still looking for the source of the Nile in northwest Africa (for the theory, see Pliny, *Natural History* 5.51–4; Ammianus Marcellinus 22.15.8; Duane W. Roller, *The World of Juba II and Kleopatra Selene* [London 2003] 192–6).

To describe Mesopotamia as a "boat" seems a misunderstanding of Eratosthenes' actual image of a rower's cushion (F83); Strabo knew better, so some intermediate source has intervened. For Zeugma see F88. On this passage, see Nicola Biffi, *Il medio oriente di Strabone* (Bari 2002) 163–4.

F88. *Summary*: **From Tomisa on the Euphrates to India are a number of places, lying on a straight line.**

Commentary: There may be little of Eratosthenes in this fragment other than the single distance cited. The passage is a typical example of Strabo's confusing layering of sources. Artemidoros of Ephesos, active at the end of the second century BC, relied heavily on Eratosthenes for his own geographical treatise, which provided an itinerary from the Euphrates to India. But the toponyms seem to reflect a later era than Eratosthenes, probably after the establishment of the independent kingdom of Kommagene around 163 BC, including Zeugma (the Seleukid crossing point rather than Thapsakos, the one used by Alexander: see F63; Getzel M. Cohen, *The Hellenistic Settlements in Syria, the Red Sea Basin, and North Africa* [Berkeley 2006] 190–6), the Kommagenean capital at Samosata, and Tomisa, which seems to have been of little importance before the second century BC (Strabo 12.2.1). Moreover, there is a density of toponyms in a region that Eratosthenes, by his own admission (F83–4), knew little about. It is probable that his distances were inserted into an itinerary of the latter second century BC.

F89. *Summary*: **The Tigris (whose name means "arrow" in Median) flows from the mountains and contains many types of fish. Mesopotamia lies on its right (and on the left of the Euphrates).**

Commentary: Like F87, this describes the course of the Tigris. Eratosthenes is only mentioned obliquely but parallelism with F87 indicates he is the source. Niphates is the ridge south of Lake Van where the source of the Tigris actually is, although the term may also have referred to the high mountains northeast of the lake (modern Tendürek Dağ, elevation 3,353 m.), where the Tigris would have originated if it flowed through the lake (Ronald Syme, *Anatolica: Studies in Strabo* [ed. Anthony Birley, Oxford 1995] 27–38). The etymology of the word Tigris was commonly mentioned in Greek and Latin sources (Pliny, *Natural History* 6.127; Quintus Curtius 4.9.16; Varro, *lingua Latina* 5.100); the meaning "arrow" seems to have some authority (Strabo, *Géographie* 8 [ed. François Lasserre, Paris 2003] 125). The single fish found in Lake Van is the darekh, a type of herring, which in recent times has reached near-mythic status. Something may have dropped out of the text because Chalonitis is in Assyria (Strabo 16.1.1, although implying the existence of another place with the same name). For the Wall of Semiramis and Opis see F83.

F90. *Summary*: **A large amount of asphalt is produced in Babylonia, both of the liquid and dry kind.**

Commentary: This is one of the more important ancient reports on asphalt. Its production in Mesopotamia and the southern Levant went back to earliest times: Noah used it in the ark (Genesis 6.14). Greeks first became aware of it through Herodotos' description of Babylon (1.179). Naphtha (Akkadian *naptu*) does not seem to have been known to Greeks until the time of Alexander, who saw a fountain of it in Mesopotamia (probably the same one cited by Eratosthenes) and was dangerously intrigued (Plutarch, *Alexander* 35), although there was a tradition that Medea had used naphtha to create the burning robe that disposed of her rival at Corinth. Perhaps not astonishingly, Daniel encountered naphtha in the burning fiery furnace (Daniel 3.46), the earliest extant reference in Greek. See also Kidd, *Commentary* 829; R. J. Forbes, *Studies in Ancient Technology* (second edition, Leiden 1964) vol. 1, pp. 1–124, especially pp. 32–42.

F91. *Summary*: **The Assyrikoi and Illyres are mentioned.**

Commentary: Stephanos of Byzantion wrote his lengthy *Ethnika* in the sixth century AC, a list of toponyms and the appropriate toponymic adjectives, which only survives in a late summary. There is rarely any elaboration, but the work is a treasure-trove, although with the toponyms usually in isolation and lacking any geographical data: the sparse nature of the entries can make identification of the names difficult. Nevertheless they provide evidence that Eratosthenes discussed a particular locality (see also F122–4, 137, 142–4, 147), and this fragment is his only extant citation of Assyria, presumably falling somewhere in the discussion of the northern part of the third sealstone (perhaps in the possible lacuna of F89).

F92. *Summary*: **The fourth sealstone is Arabia Eudaimon, the Arabian Gulf, Egypt, and Aithiopia.**

Commentary: Having had serious difficulties with his third sealstone, Eratosthenes made a valiant try for a fourth, consisting of the Arabian peninsula, Egypt, and Aithiopia. Rather than the complex measurements and distances used for the third sealstone, there is an astonishingly dogmatic definition of the territory as bounded by two meridians and two parallels, returning, in theory, to the quadrilateral shape that he seems to have believed was appropriate. Either Eratosthenes did not wish to plunge again into the complexities of defining boundaries along known routes with known distances, or such data have not survived in Strabo's version. The extant material on the fourth sealstone is more cultural than geographical, and is the major extant source for pre-Roman Arabia and the Persian Gulf. The phrase "Arabia Eudaimon" ("Happy Arabia") refers to the southwest corner of the Arabian peninsula (near Aden), and became common in early Hellenistic times before there was direct trade between the Mediterranean and India. Happy Arabia was the transfer point between Mediterranean and Indian commodities and thus was fortunate in that it received both (*Periplous of the Erythraian Sea* 26). The Roman translation, Arabia Felix, has a slightly different meaning, referring to the abundance of aromatics available in the region (first in Pomponius Mela [3.79, as Arabia Eudaemon], and then Pliny, *Natural History* 5.87, as Arabia Felix).

F93. *Summary*: **The Erythraian Sea has two gulfs, the Persian on the east and the Arabian on the west, with Arabia between them.**

Commentary: Pliny preserved Eratosthenes' circumference of the Persian Gulf, translated to 2,500 miles (compare 10,000 stadia for the north side in F94). The source for these numbers was the cruise of Androsthenes of Thasos (for whom, see F94). Pliny's "second bay" or Arabian Gulf is the modern Red Sea, but in Eratosthenes' day the outer coast of Arabia between the two gulfs had not been explored. Alexander had commissioned Hieron of Soloi to go from the Persian Gulf around Arabia into the Red Sea, but he lost his nerve and turned back shortly after leaving the gulf (Arrian, *Anabasis* 7.20.7–8), and it was not until the Augustan period that the coast of the peninsula was known in its entirety, first described by Juba II of Mauretania (*FGrHist* #275, F30–3) Hence Pliny's figure for the total extent of Arabia cannot be from Eratosthenes. Moreover, the term "Azanian" (the western part of the Indian Ocean and the East African coast) seems not to be pre-Roman: it was unknown to Strabo but used regularly by the author of the *Periplous of the Erythraian Sea* (15–16, 31, 61: on the toponym, see John Hilton, "Azania—Some Etymological Considerations," *AClass* 35 [1992] 151–9). Yet Eratosthenes is probably one of the unnamed sources who provided further dimensions of the Persian Gulf and perhaps the image of its shape.

F94. *Summary*: **The mouth of the Persian Gulf is so narrow that one can look across it. From its mouth it turns to the north and extends about 10,000 stadia. Information about the gulf is supplied by Androsthenes of Thasos, who sailed around it, reporting on the city of Gerrha and the islands of Tyros and Arados. Nearchos and Orthagoras described the island of Ogyris, 2,000 stadia out to sea. The coast of the region is noted for its peculiar trees.**

Commentary: Eratosthenes' thorough report on the Persian Gulf is based on the treatise of Androsthenes of Thasos (*FGrHist* #711): in fact, this is the longest surviving fragment of Androsthenes' work (his F2), and Eratosthenes may have been the last to see a copy of it. Androsthenes was a companion of Alexander who returned by sea from India under the command of Nearchos. He was commissioned by Alexander to explore the Persian Gulf, in the winter of 325/4 BC or the following one, summarizing his results in a brief report, *Sailing Along the Indian [Ocean] Coast*. The first part of Eratosthenes' summary is a description of sailing into the Persian Gulf from India, around the long northward-pointing promontory of Arabia and into the narrow Straits of Hormuz (ancient Harmozai). The distance of 10,000 stadia is quite reasonable, as is the comparison with the Euxine. With mention of the important city of Teredon, the account changes from the movement up the gulf to Androsthenes' actual journey. Teredon, the probable starting point for his cruise, lay near when the Euphrates and Tigris joined, perhaps near modern Basra (F83). The first place he reached was the island of Ikaros, modern Failaka off the Kuwaiti coast, which had been given its Greek name by Alexander (Arrian, *Anabasis* 7.20), based on an indigenous name for the temple on the island, Ichara (Ptolemy, *Geography* 7.4.47): the temple cult could be assimilated with Apollo and Artemis Tauropolos. From Ikaros, Androsthenes sailed to the trading city of Gerrha, whose location is not certain, but perhaps near modern al-Jubayl in Saudi Arabia (see Daniel T. Potts, "Thaj and the Location of Gerrha," *PSAS* 14 [1984] 87–91). It had long existed, populated by Chaldaians who had probably been exiled from Babylonia in the Persian period. Androsthenes was intrigued by their salt-based architecture. There is also a passing reference to Aristoboulos of Kassandreia (*FGrHist* #139), who wrote a history of Alexander (see also F109), perhaps consulted directly by Eratosthenes or Strabo. Androsthenes next went to the island of Tylos, or Tyros, modern Bahrain, where he made an extended visit. The alleged Phoenician connection with the Persian Gulf was ancient (Herodotos 7.89; also Strabo 16.4.27), and there is little reason to doubt the tradition of an early Phoenician migration to the region (see G. W. Bowersock, "Tylos and Tyre: Bahrain in the Graeco-Roman World," in *Bahrain through the Ages: The Archaeology* [ed. Shaikha Haya Ali Al Khalifa and Michael Rice, London 1986] 400–2).

The comments about the mysterious island of Ogyris may have been included in Androsthenes' report, or may be Eratosthenes' insertion, as the sources are Nearchos (*FGrHist* #133) and Onesikratos (*FGrHist* #134) and the island is out to sea, so would have been encountered before the fleet reached the Persian Gulf. The text actually reads Τυρίνην rather than Ὠγυρίν, so either Strabo or Eratosthenes misread his source: it is clear that Ogyris is meant by its location (2,000 stadia out to sea, nowhere near Tyros) and the parallel passage by Juba II (*FGrHist* #275, F31). Despite the precise information, it cannot be located. The account returns to the Persian Gulf with a description of the peculiar flora of the region, presumably mangroves, described in greater detail in Androsthenes' F4. On the material in this passage, see D. T. Potts, *The Arabian Gulf in Antiquity* 2: *From Alexander the Great to the Coming of Islam* (Oxford 1990) 154–96.

F95. *Summary*: **The northern part of Arabia is desert; the southern is called Eudaimon, which is fertile and productive. There are many ethnic groups, with several prosperous monarchies. Frankincense and myrrh are major commodities, brought to coastal ports. To the west of Arabia is the Arabian Gulf, opposite the Trogodytic territory. The gulf narrows at Deire, where there is a pillar of Sesostris, who crossed into Asia. No one has gone beyond the mouth of the gulf.**

Commentary: Eratosthenes' account of Arabia is the earliest extant. The peninsula was hardly known to the Greek world before the time of Alexander, who had considered conquering it (one erroneous tradition even said he had done so: see Pliny, *Natural History* 12.62). Yet Arabian aromatics reached the Makedonian court (Plutarch, *Alexander* 25.4–5; *Sayings of Kings and Commanders* [179e–f]), which stimulated interest in the peninsula. Aristotle's student Palaiphatos of Abydos (*FGrHist* #44) seems to have written the first treatise specifically on Arabia, but there is no record of any other detailed study before Eratosthenes. The single source mentioned in the passage, and significantly not about Arabia proper, is Anaxikrates (the only citation of his name), whom Alexander commissioned to explore the Red Sea and who provided some of the first data about its length and coasts, and was perhaps the first to visit the frankincense territory. In addition to the material summarized by Eratosthenes, two other accounts of this cruise survive (Theophrastos, *Research on Plants* 9.4.4; Arrian, *Indika* 43.7), but neither mentions Anaxikrates by name (see W. W. Tarn, "Ptolemy II and Arabia," *JEA* 15 [1929] 14–16; Paul T. Keyser, "Anaxikrates," in *The Encyclopedia of Ancient Natural Scientists* [ed. Paul T. Keyser and Georgia L. Irby-Massie, London 2008] 74).

There would have been some material available on Arabia from the Alexander sources, but what Eratosthenes reported goes far beyond the existing coastal reports, and was probably obtained from traders and merchants in Alexandria. The signatures of information received in this manner are apparent: a focus on the great trading centers of Petra and Gaza, descriptions of the difficulties of overland travel, emphasis on the various populations one would pass through, use of days as a distance, and specific references to trade. The account is geographically linear, moving from north to south, following the trade route from Petra (which is connected to Babylon and thus Eratosthenes' third sealstone) to the aromatic-producing region of the southwest. There is a brief mention of the wedge of territory north of Arabia and west of Mesopotamia (Judaea and Syria), not otherwise discussed in the extant fragments except for the enigmatic F18 and 96, and, as noted, strangely left out of the third sealstone (see F82). For Arabia Eudaimon see F92. The account of Arabia proper begins at Petra, which was just beginning to come into its own as an important center, having been able to establish itself as an independent state in the time of the Successors. By Eratosthenes' day it had become a focal point for trade with Arabia, at the junction of routes to the Mediterranean at Gaza, Mesopotamia, and Arabia itself. Eratosthenes' description follows this last route south into the heart of the peninsula, describing the change from farming territory to a more barren and hostile land suitable only for camel nomads. The account then jumps to the coast of the Red Sea (modern Yemen) with an exaggerated description of Indian-like lushness, perhaps reflecting the relief of the caravans when they reached this area, the most fertile part of Arabia, with an agricultural prosperity not apparent elsewhere. The focus is the four major ethnic groups, described roughly east to west, as they were probably encountered coming from Petra. Mariaba may be modern Marib (but see A.F.L. Beeston, "Some Observations on Greek and Latin Data Relating to South Arabia, *BSOAS* 42 [1979] 10), and Sabata modern Shabwa. Other and later sources for the region have similar information: the most important are *Periplous of the Erythraian Sea* 27; Pliny, *Natural History* 6.154–5 (based on the *On Arabia* of Juba II), and Ptolemy, *Geography* 6.7.38. The description has an autopsic character, and the peculiar ethnographic note on the method of succession sounds like taverna gossip. For the cultures and environment described, see the catalogue *Queen of Sheba: Treasures from Ancient Yemen* (ed. St. John Simpson, London 2002).

The most notable feature of this region was its the production of the aromatics frankincense (*Boswellia sacra*) and myrrh (*Commiphora myrrha* Nees), which Eratosthenes mentioned only in passing (the most complete descriptions are those of Theophrastos [*Research on Plants* 9.4] and Juba II, preserved in Pliny, *Natural History* 12.51–81). Although the trade had not reached the large levels it would in the Augustan period, in Eratosthenes' time it was beginning to become important, due in part to the legacy of Alexander. Frankincense and myrrh were often mentioned together (e.g. the famous passage in Matthew [2.11]), although myrrh came from a wider area and had been known since early times (Genesis 37.25); frankincense was first cited by Herodotos (1.183). For the frankincense and myrrh trade generally, see Gus W. Van Beek, "Frankincense and Myrrh," *BiblArch* 23 (1960) 69–75. Eratosthenes then mentioned another trade route into the region, the 40-day trip from Gerrha on the upper Persian Gulf (F94), which leads to a digression on exploration of the coasts of the peninsula, especially the Red Sea side, based on the travels of Anaxikrates (supra, p. 195), who went from Heroonpolis (the Gulf of Suez) to the mouth of the sea at Deire ("The Neck"), at the Bab el-Mandeb (see Lionel Casson, *The Periplus Maris Erythraei* [Princeton 1989] 116). It is probable that Anaxikrates was to connect with Hieron of Soloi, coming around Arabia clockwise, but this never happened. Anaxikrates continued along the coast of the Kinnamomon Bearers, the Horn of Africa, but no farther; despite the use of νυνί and νῦν, the passage cannot refer to Strabo's own day, when the circumnavigation of Arabia had been accomplished and there was knowledge of the East African coast. Juba's *On Arabia*, written no later than AD 5, is the first to provide the distance around Arabia (*FGrHist* #275, F33) and a distance of 1,875 miles south along the African coast, to the vicinity of Zanzibar (his F35; see Duane W. Roller, *The World of Juba II and Kleopatra Selene* [London 2003] 227, 242). Moreover, even in Eratosthenes' time there was information beyond the mouth of the Red Sea, from Timosthenes of Rhodes (Pliny, *Natural History* 6.198), a source that Eratosthenes knew (F134), so the present passage closely reflects the earlier perspective of Anaxikrates.

A final digression is the artifact of Sesostris that Anaxikrates reported from Deire. Herodotos (2.102–11, 137) was the first to describe the exploits of this legendary Egyptian king; his account is supplemented by Diodoros (1.53–8). Although there were three kings of the Twelfth Dynasty named Sesostris (1956–1852 BC), the Greek tradition, if it ever had any connection with these rulers, had developed by Classical times into a romantic view of the great conqueror. He allegedly subdued peoples on the Red Sea and in Aithiopia and then crossed over to Arabia and was victorious everywhere, eventually reaching as far as Thrace, Skythia, Kolchis, and even India, setting up memorials of himself wherever he went. Upon his return to Egypt he became a great builder and social reformer. The data are a complex amalgam of actual activities of Egyptian kings and the mythical great conqueror who became an Egyptian prototype for Alexander that he could eventually surpass. Even by the time of Herodotos there was a tendency to attribute monumental reliefs in western Asia to him: the ones that Herodotos saw in Anatolia were probably Hittite. Exactly what Anaxikrates saw at Deire remains uncertain.

F96. *Summary*: **There are lakes near Arabia from which water flows underground as far as Koile Syria.**

Commentary: Strabo's context is Mesopotamia and its hydraulic phenomena, especially Alexander's interest in them, but the passage is inscrutable and Strabo himself did not seem to understand it. There are no known lakes that seem to fit the location: Strabo believed it was the marshes and flooded lowlands of the lower Euphrates. Any connection, however, with Koile ("Hollow") Syria is dubious, whether the toponym is meant in its localized sense of the upper Orontes valley (generally over 1,000 m. above sea level) or its more general usage of inland Syria, all areas over 1,000 km. from the lower Euphrates. Inclusion of the territory around Rhinokoloura and Mt. Kasion (the eastern Egyptian coast) only complicates matters. The passage may reflect some knowledge of the Jordan rift valley and the Dead Sea, far below sea level, and takes for granted the persistent ancient theory of underground rivers connecting watercourses far apart (see F87, 140), but it involves a region (the Levant) hardly discussed in the extant fragments, and has become hopelessly confused. See Nicola Biffi, *Il Medio oriente di Strabone* [Bari 2002] 149.

F97. *Summary*: **The Persian Gulf is 1,200 miles long.**

Commentary: Pliny listed three sources for the length of the Persian Gulf. The earliest is Timosthenes of Rhodes, who was sent by Ptolemaios II to look for harbors on the Red Sea (Strabo 9.3.10), and who also had a career as a famous composer. His report, *On Harbors*, also included the Mediterranean, and was ambivalently received by Eratosthenes (see F131, 134, 140). Timosthenes provided sailing days for the perimeter of the sea (except the narrows, fairly easily estimated), which Eratosthenes converted to stadia and Pliny to miles. For Artemidoros see F88.

F98. *Summary*: **The Nile is about 1,000 stadia west of the Arabian Gulf, and flows north from Meroë, which is on an island. There are a variety of peoples along the river, including the Egyptian fugitives above Meroë. The Trogodytes are 10–12 days' journey from the Nile.**

Commentary: Since Eratosthenes considered Egypt part of the fourth sealstone along with Arabia, he moved smoothly from the peninsula across the Red Sea and into Egypt proper, although the term "sealstone" has quietly vanished from his terminology, never to return. As an introduction to Eratosthenes' account of the Nile, Strabo has included some structural details of his own *Geography* that do not apply to Eratosthenes' work.

Much of the material about the Nile had already appeared, in somewhat fragmented form, in Eratosthenes' Book 2 when he outlined his prime meridian. It is not always possible to determine how repetitive he might have been or how Strabo broke up his arguments to fit his own work. The passage is an account of the Nile from the Island of the Fugitive Egyptians (see F35) to Syene, with emphasis on the great bends of the upper river, especially the western loop between Meroë and Syene. No sources are mentioned, but presumably the primary ones were the explorers of Ptolemaios II such as Philon (F40), Dalion, and Simonides. Distances from Meroë north are cited, much more precise (and somewhat larger) than the meridian distances recorded in F58, demonstrating that Eratosthenes did make allowances on occasion for the differences between measured land distances and those along a meridian. Several tributaries are mentioned, all with the prefix "asta." The passage is somewhat confused, as there are two localities called Meroë, one an island at the junction of major affluents, and the other 700 stadia to the north. Problems with the several river toponyms have aggravated the difficulties. Lower Meroë is probably the island where the modern Atbara joins, and upper Meroë the traditional locale of the famous city. The account up the river continues to the Island of Fugitive Egyptians, two months farther according to Herodotos (2.30–1). The southernmost tributary, the Astapous, although placed below the city of Meroë, sounds like the White Nile, originating in lakes to the south. Meroë was allegedly ruled by a dynasty of queens titled Kandake (Strabo 17.1.54), which seems somehow to have been transferred to the Island. The passage closes with a catalogue of ethnic groups on either side of the river, including the Trogodytes (see F41) on the coast and the Nubians on the left side of the river.

F99. *Summary*: **The Nile floods because of the rains.**

Commentary: Proklos' commentaries on Plato's works were written in the fifth century AC. He is the only source to report that Eratosthenes mentioned the flooding of the Nile. Although one must be dubious about data that first appear in late commentaries, it is improbable that any discussion of the river could have avoided its most persistent intellectual problem, an issue as old as Thales of Miletos (Aetios 4.1.1.).

Eratosthenes' theory goes back to Herodotos (2.20–5), and there is no obvious reason why Proklos singled him out.

F100. *Summary*: **Libya is smaller than Europe, and of lesser importance. Much of it is uninhabitable, except for the Mediterranean coast. It is shaped like a right-angle triangle. If one goes beyond the Pillars of Herakles one reaches the small town of Lixos, opposite Gadeira, and beyond are many Phoenician mercantile settlements.**

Commentary: Strabo's summary of Libya (Africa west of Egypt) only mentions Eratosthenes near the end, but clearly dates from his time. Strabo shows no knowledge of the *Libyka* of Juba II of Mauretania, published before 2 BC while Strabo was still working on his *Geography* (Duane W. Roller, *The World of Juba II and Kleopatra Selene* [London 2003] 183), and which would have superseded much of this information. Strabo's summary also predates the fall of Carthage in 146 BC, and thus is earlier than Artemidoros, the only other source mentioned in the passage.

Libya was believed to be in the shape of a right-angle triangle, with its hypotenuse running from Aithiopia to the Atlantic coast. This is an ancient concept of Africa: Hesiod (*Catalogue*, F45) believed that the continent was exceedingly narrow. Despite the early circumnavigations, there was no understanding of the southern extent of the continent: it was thought that one could go directly from the Somali coast to West Africa. Eratosthenes, concluding that most of the continent was inaccessible, conjectured its size, using his previously calculated distances along the Nile for the east side, and presuming that the north side was twice that distance. There is no mention of a sealstone. The description of the Mediterranean coast is astonishingly skimpy, perhaps because in Eratosthenes' day everything beyond Kyrene was under Carthaginian control. If he devoted any particular care to his home territory of Kyrene, it has not survived.

The greatest detail is reserved for the Atlantic coast beyond the Pillars of Herakles, the western part of the district called Maurousia (Mauretania), in Eratosthenes' day ruled by an indigenous kingdom dependent on Carthage. Eratosthenes may have been the first to describe Maurousia, probably using the account of a Greek traveler who visited the Atlantic coast in the middle of the fourth century BC. His name is not known, but a summary of his report survives in the text attributed erroneously to Skylax (*GGM* vol. 1, pp. 15–96). This traveler mentioned many of the same toponyms, including Lixos, the most important city in coastal northwest Africa, whose name survives in the Leucos River of Morocco. The coast was dotted with Carthaginian trading posts, many of which went back to the reconnaissance of Hanno around 500 BC. Yet the Greek account shows no knowledge of the farther part of coastal West Africa, beyond the desert, which was reached not only by Hanno but his contemporary Euthymenes of Massalia. There is knowledge of the Atlas mountains (first mentioned by Herodotos, 4.184), conceived, in the traditional fashion of Eratosthenes, as a lengthy east-west range running all the way to the environs of Carthage at the Syrtos Gulf. There is also the earliest mention of the Gaitoulai, a generic term for the interior indigenous peoples of northern Africa, who were to cause great difficulty for the Romans in the second and first centuries BC (Sallust, *Jugurtha* 18). For Eratosthenes and Maurousia, see El Mostafa Moulay Rchid, "Ératosthène. L'*Oikoumène* et la Maurusie: Géographie et symétrie," in *Mélanges Pierre Lévêque* 3 (Paris 1989) 269–75.

F101–3. *Summaries*: **F101: It is 525 miles from Kyrene to Alexandria by land. F102: It is over 13,000 stadia from Alexandria to Karchedon. F103: It is 1,100 miles from the Ocean to Carthage, and 1,628 miles farther to the Kanopic Mouth.**

Commentary: Many distances published by Pliny were derived from Eratosthenes. Generally these are listed without additional details, often in a comparative list from a range of Greek and Roman sources. They are almost always converted into miles, which can occasionally be compared with known stadia distances over the same route. F101 and 103 (which are only separated in the *Natural History* by a brief reference to the map of Marcus Agrippa) are in a long catalogue of distances along the north coast of Africa: Eratosthenes is the earliest source mentioned. F102 (actually part of F65, Eratosthenes' discussion of the meridians) provides a comparison with F103, assuming that both are measurements from Carthage to Alexandria (which is a few kilometers west of the Kanobic mouth), and provides a calculation of eight Roman miles (11.8 km.) to a stadion, although the distance in F102 is rounded.

F104. *Summary*: **The Great Syrtes is 5,000 stadia across and 1,800 deep.**

Commentary: The Syrtes are the southern bays of the Mediterranean that lie between the Kyrenaika and Carthage, with the Lesser Syrtes to the west. Automala, the western boundary of the Kyrenaika, is the southernmost point of the Mediterranean; Meninx (see further, F105) and Kerkina are the islands off the coast of Tunisia. This sparse account is part of Strabo's survey of the coast of the Mediterranean (2.5.17–25). Except for a somewhat irrelevant citation of Demosthenes at 17, Eratosthenes is the only source mentioned (again at 24 = F128), so he may have provided most of the information for the general survey, but this cannot be proven.

F105. *Summary*: **There are few islands off the northern Libyan coast, but the most notable is Lotophagitis.**

Commentary: Pliny's description of the islands off the north African coast attributes to Eratosthenes the statement that Meninx (modern Jerba in Tunisia) was the island of the Lotos Eaters. Other data in this passage may also be from Eratosthenes, although the comment about scorpions is a more Roman point of view (see Vitruvius 8.3.24; Lucan 9.294–949). The land of the Lotos Eaters may have been discussed by Eratosthenes in his Homeric critique of Book 1. Its location had long been a point of speculation, especially since it was the first place Odysseus encountered after the storm that kept him from rounding the southern end of Greece (*Odyssey* 9.82–104). Thus the Lotos Eaters were an essential connection between the known and unknown world and the key to the topography of the wanderings. Odysseus was blown by a north wind past the island of Kythera, south of the Peloponnesos, with the storm continuing for nine days, so North Africa would be the probable region for the Lotos Eaters. The first to be specific was Herodotos (4.177) placing them on a peninsula occupied by the Gindarnes, somewhere west of Kyrene. There are no peninsulas on this portion of the North African coast except the two adjacent ones at Zarzis, in the eastern part of modern Tunisia. Just offshore is the island of Jerba, ancient Meninx, Eratosthenes' location for the Lotos Eaters, and now, not unexpectedly, a major tropical resort. Whether he was originally responsible for the identification is uncertain, but it was taken for granted, without attribution, by Strabo (17.3.17).

F106. *Summary*: **There are varying locations for the Pillars of Herakles, but most believe they are at the narrows. The Libyan pillar is Abilyx, in the territory of the Metagonioi.**

Commentary: The exact location of the Pillars of Herakles was long a matter of dispute. Although they may seem obvious today as the two large mountains at the western end of the Mediterranean, Gibraltar and Jebel Mousa, such was not the case in antiquity, and understanding of the region changed as topographical knowledge increased. At some early date, Homer's mythical and unlocated Pillars of Atlas (*Odyssey* 1.51–4) became associated with the wanderer Herakles, but as the western end of the Mediterranean became better understood in the latter seventh century BC, uncertainly increased rather than decreased. Herodotos, who mentioned the Pillars several times, placed them east of Gadeira and Tartessos (4.8, 152), which could mean anywhere in the 50-kilometer-long strait (the modern Strait of Gibraltar) that runs east to the opening of the Mediterranean, through rugged topography with several promontories that could be identified as the Pillars, although especially prominent are Gibraltar and Jebel Mousa (the Kalpe and Abilyx of Strabo) at the eastern end. The early prominence of Gadeira caused some (such as Pindar) to place them in that area, or at points east thereof, such as Tarifa or Cape Trafalgar: the sources seem uncertain as to whether height or prominence was the defining criterion. It was Dikaiarchos (his F125) who seems to have definitely located them at Gibraltar, a view canonized by Eratosthenes.

F107. *Summary*: **Lixos is at the western extremity of Maurousia. There are a large number of destroyed Phoenician settlements in this region, and the air is brackish and misty.**

Commentary: Presuming that the "because" (διότι) refers to the entire passage, Eratosthenes provided some detailed information about the Atlantic coast of Africa, derived in all probability from the anonymous Greek traveler of the fourth century BC (see F100). The description of coastal fog, something largely alien to the Mediterranean, is vivid and seems an eyewitness account. The many deserted Carthaginian cities (300, according to Strabo 17.3.3) shows the decline of Carthaginian power and may even be a fact from Eratosthenes' own era of the Punic Wars.

F108. *Summary*: **There are a large number of ethnic groups to the northwest of India, along the Iaxartes and Oxos, stretching as far as the Skythians. It is 22,670 stadia from the Kaspios to the Iaxartes, and 15,300 from the Kaspian Gates to India.**

Commentary: The northeastern part of the inhabited world—the region from the eastern Black Sea through the Caucasus to the Caspian Sea and beyond—was poorly understood in Eratosthenes' time. Greeks had

first learned about the area in connection with the final campaign of
Cyrus the Great in 530 BC, among the Massagetai far to the east of the
Caspian Sea, reported by Herodotos (1.201–10). The Caucasus region
was also early associated with Prometheus (Aischylos, *Prometheus* 707–
35). But it remained at the fringes of Greek knowledge until the time of
Alexander, whose companions manipulated the topography of the area
in order to enhance his reputation (F23–4), something that hardly assisted geographical understanding.

No sources are mentioned, but the most probable is Patrokles, the
Seleukid explorer of the region (F47), and one or more of the road itineraries used by Eratosthenes (F78, 81), perhaps in part going back to
Alexander's surveyor Baiton (*FGrHist* #119). The first section of the
fragment is an ethnography of the Caspian Sea region and points east.
A number of these peoples lay east of the sea and had been encountered
by Cyrus. These were in the vicinity of the Oxos and Iaxartes Rivers, the
modern Amu Darya and Syr Darya of central Asia. Another group is the
peoples on the shores of the Caspian Sea, of whom little was known, and
then there are those west of the sea and close to the Armenian mountain of Parachoathras. The list from the Caspian Sea coast is almost
certainly from Patrokles, who was the first to explore it (Strabo 2.1.17).

The second part of the fragment is two itineraries. The first is from
Mt. Kaspios south of the Caspian Sea to the Iaxartes, not following the
wandering route of Alexander but the most direct line between the two
points, ending at the site of Alexandria Eschate (modern Khojend in Ta-
jikistan), Alexander's most remote settlement, founded in 329 BC (Arrian, *Anabasis* 4.1.3). The second itinerary, from the Kaspian Gates to
the borders of India, follows Alexander's route more closely (without its
northern detour). Both represent ancient trade routes that were used
before and after the time of Alexander.

F109. *Summary*: **Patrokles reported that the rivers through Hyrkania are the Oxos and Ochos, emptying into the sea. The Oxos
is the largest Asian river outside India, and is navigable. Indian
commodities come down it and eventually end up at seaports on
the Euxine.**

Commentary: The rivers of the extreme northeast were uncertainly defined. The Oxos and Ochos may represent the same indigenous name,
although the sources are clear in distinguishing them. The Oxos was the
larger and better known, the modern Amu Darya, flowing from the gla-
ciers of the Hindu Kush through Baktria to the Aral Sea; its Baktrian

portions became known at the time of Alexander. The Ochos, often mentioned along with the Oxos (F24), is difficult to find: Eratosthenes made it clear that it was a major river, and it must lie south of the Oxos because Alexander crossed the Ochos first (Quintus Curtius 7.10.15), but the Oxos / Amu Darya has few southern tributaries. The best suggestion is the modern Kunduz, but the matter remains disputable (Bosworth, *Commentary* vol. 2, p. 110). Aristoboulos of Kassandreia (*FGrHist* #139), the companion and historian of Alexander, was used frequently by Strabo for India, but this is the only place he is mentioned in conjunction with Eratosthenes, although it is not obvious that the latter used him. The Oxos is indeed the largest river of central Asia (2,500 km. long), and the lower parts remain navigable despite heavy use for irrigation (Walbank, *Commentary* vol. 2, pp. 262–3). However, it does not flow into the Hyrkanian (Caspian) Sea, but empties into the Aral Sea. The confusion was probably due in part to the manipulation of topography by Alexander's companions, as well as the remoteness of the territory, as it was generally thought that all major rivers of central Asia emptied into the Caspian Sea. There is no doubt that the Oxos was part of a significant trade route that eventually went across the Caucasus to the Black Sea, and there may have been a path from the Aral Sea across the Ust-Urt plateau to the Caspian. The Aral Sea was barely known, if at all, in Hellenistic times: see J. R. Hamilton, "Alexander and the Aral," *CQ* n.s. 21 (1971) 106–11. Patrokles' suggestion that Indian goods came down the Oxos to the Caspian, despite the problems of topographical detail, is quite reasonable, although somewhat at odds with his notorious report that India and the Caspian were connected by the Ocean (Strabo 11.11.6).

The Caspian Sea was much larger in antiquity than today (S. N. Mouraviev, "Ptolemy's Map of Caucasian Albania and the Level of the Caspian Sea, *VDI* 163 [1983] 117–47), and thus the routes and river systems may have been different. F109 also defines the way trade goods went from the Caspian to the Black Sea. The fertile valley of the Kyros River (the modern Kura, which preserves the ancient name) provides a gentle access to the interior from the southwestern Caspian (in modern Azerbaijan south of Baku). At the source of the river (above modern Tbilisi) are a number of low passes (none more than 1,000 m. above sea level) that lead to the source of the Phasis (modern Rioni) and eventually Kolchis. Given that this route runs less than 75 km. south of the Kaukasos Mountains, whose summits exceed 5,000 m., it is a remarkably easy way through rugged territory, used since prehistoric times.

F110. *Summary*: **The Kaspian Sea is an inlet of the Ocean, extending to the south; its entire circuit is known to the Greeks.**

Commentary: Strabo, following Eratosthenes, had divided the northern part of the world into two sections (11.1.5), beginning the second section at the Kaspian (Hyrkanian) Sea. The two names seem totally interchangeable, both originally ethnyms, with the Kaspioi to the south and the Hyrkanoi to the southeast. In Eratosthenes' day the sea was believed to be an inlet of the Ocean. Herodotos (1.203) and Aristotle (*Meteorologia* 2.1) knew otherwise, but the readjustments of local topography at the time of Alexander, as well as his desire to build a fleet on the sea to reach the western Mediterranean (Arrian, *Anabasis* 7.15–16), resulted in the conventional wisdom that the sea was part of the Ocean, a view that lasted until medieval times (see *Commentary* to F24). It was even said, improbably, that this theory had existed long before Alexander (Plutarch, *Alexander* 44). If there were any topographical basis for this belief it may have been due to encountering a river flowing into it from the north (perhaps the Volga) that might have seemed a connection to the Ocean (Pomponius Mela 3.38). Although Herodotos (1.203) had some information about the dimensions of the sea, expressed in sailing days, the circuit, measurements, and ethnic data known to Eratosthenes probably came from Patrokles, who may have identified the two large rivers flowing into its north end, the Volga and Ural, as the Oxos and Iaxartes, both of which in fact only reached the Aral Sea.

F111. *Summary*: **The extent of the Kaspian Sea from the Kadousian coast to the Iaxartos is 16,300 stadia.**

Commentary: Pliny reported Eratosthenes' measurements (from Patrokles) for the Caspian Sea (as in F110), but the one along the Albanian (west) coast has nearly doubled. In addition, the Oxos River has been replaced by the Zonos, otherwise unknown. The total distance Pliny reported is 16,300 stadia, which would give 11.35 miles to the stadion, an unlikely equation (compare F103, the more normal eight miles). Yet if the 9,000 stadia of the Albanian coast were corrected to the 5,400 of F110, the conversion becomes exactly eight to the mile, proof of the error in Pliny's text.

F112. *Summary*: **The Kaukasiai live near the Kaspian Sea.**

Commentary: The scholiast to Apollonios of Rhodes provided several details about the northern part of the inhabited world not cited elsewhere

in the extant fragments of Eratosthenes (see also F120–1, 145, 148–9).
As is often the case with scholia, there is little context or detail. The
ethnym Kaukasiai is not otherwise known in the surviving fragments,
but to Strabo (11.2.16) it described the peoples around Dioskourias at
the east end of the Black Sea, a region discussed by Eratosthenes (F52),
and thus probably the context of F112.

F113. *Summary*: **The Kaukasos is called the Kaspian by the locals.**

Commentary: The similarity of the names Kaukasos and Kaspian (and
their variants) often caused confusion, and they were often mentioned
in tandem (Herodotos 1.203).

F114. *Summary*: **The circuit of the Pontos is 23,000 stadia and it
is shaped like a Skythian bow.**

Commentary: There are several extant distances for the circuit of the
Black Sea, all between 20,000 and 25,000 stadia, but this is the only citation that mentions Eratosthenes. Ammianus claimed that the distance went back to Hekataios, without specifying whether he meant
the Milesian (*FGrHist* #1) or Abderan (*FGrHist* #264), although both
predated Eratosthenes. The imagery of the Skythian bow appears at
Strabo 2.5.22, although no source is cited: Ammianus explained it
(22.8.37) as a straight center with long curved ends.

F115–16. *Summaries*: **F115: From the Bosporos to the Maiotic
Lake is 1,338½ miles. F116: From the mouth of the Pontos to the
mouth of the Maiotis is 1,545 miles.**

Commentary: In two separate places Pliny recorded Eratosthenes' distance across the Black Sea: the "100 less" of F115 is probably rounding.
A distance of 1,545 miles is 12,360 stadia (at eight to the mile) or 2,287
km.; the distance is actually about 725 km., so the ancient figure may
reflect a coastal sailing route.

F117. *Summary*: **The passage at the Symplegades is narrow and
crooked.**

Commentary: The toponym Symplegades ("Clashers") does not appear
before the fifth century BC, but the name seems a variant of the Homeric Planktai (*Odyssey* 12.61, 23.327), the "Wanderers." There were

also the Kyaneai ("Dark Ones"), which Herodotos (4.85) equated with the Planktai and placed at the mouth of the Bosporos. All these are typical sailors' tales and tended to move around the Mediterranean (and eventually beyond) as appropriate. Euripides made the name Symplegades popular, expectedly in *Iphigeneia among the Taurians* (260, 355, 1389) and *Medea* (1263), and also connected them with the Argonauts (*Andromache* 795), thus firmly locating them at the mouth of the Black Sea (although some in Eratosthenes' day did not agree: see F106), where there are a number of navigational hazards.

F118. *Summary*: **There are hidden and unseen rocks around the Euxine called the Synormades.**

Commentary: The *Alexandra*, attributed to Lykophron but probably of the second century BC (and thus after Eratosthenes), placed the Clashers (F117), as was customary, at the mouth of the Black Sea. Tzetzes, in his commentary of the twelfth century, said that Eratosthenes had called them the "Movers." The word συνορμάδες and its relatives are rare: other than a dubious use by Simonides (his F546), all citations seem no earlier than the Roman period, so the diction suggested by Tzetzes is doubtful.

F119. *Summary*: **The Phasis and Thermodon empty into the Pontos.**

Commentary: This fragment records an error by Eratosthenes. The Lykos River, probably the modern Kelkit Çay, empties into the Black Sea in Anatolia near Herakleia (modern Ereğli). Eratosthenes mistakenly put down the name Thermodon, probably the modern Terme Çay at Themiskyra, some distance to the east. See also F120–1.

F120–1. *Summaries*: **F120: The Phasis flows from the Armenian mountains to Kolchis. F121: The land of Titenis is named after the Titenis River.**

Commentary: Two citations from the scholia to the *Argonautika* of Apollonios of Rhodes reveal (as with F119) Eratosthenes' interest in the rivers emptying into the Black Sea. The Phasis (see also F8, 52, 119) was the major river flowing into the southeastern Black Sea, at Kolchis, the modern Rioni. It had been known to Greeks since the time of Hesiod (*Theogony* 340) and was early associated with the Argonauts (Herodotos 1.2).

The Titenis River and its territory are unknown, but the *Argonautika* (4.131–2) seems to place it along the northeastern coast of Anatolia.

F122–4. *Summaries*: **F122: Gangra is mentioned. F123: Amaxa is a district of Bithynia. F124: Tersos and Tarsenoi are mentioned.**

Commentary: Three citations from the *Ethnika* of Stephanos of Byzantion (see F91) provide Anatolian toponyms. Gangra is in Paphlagonia, modern Çankırı, and was the royal residence of the Paphagonian kingdom (Strabo 12.3.41; A. H. M. Jones, *The Cities of the Eastern Roman Provinces* [Oxford 1937] 168–9). Amaxa in Bithynia is obscure. Tersos is a variant of Tarsos, the famous city of southern Anatolia: see further, F125.

F125. *Summary*: **Tarsos is named because of Zeus Tersios.**

Commentary: Dionysios wrote his *Circuit of the Inhabited World* in the early second century AC, and Eustathios made his commentary on it a millennium later. The ancient cult of Zeus Tersios in Kilikia was said to be the source of the name of the city of Tarsos (see also F124: on the god and his cult, see Arthur Bernard Cook, *Zeus: A Study in Ancient Religion* [Cambridge 1914] vol. 1, pp. 593–604). The earliest coins from Tarsos (450–380 BC) have ΤΕΡΣΙ (and "Tarz" in Aramaic); ΤΕΡΣΙΚΟΝ was also used in the fourth century BC (Barclay V. Head, *Historia Numorum* [new and enlarged edition, London 1911] 729–31). The name "Tersos" was probably still current in Eratosthenes' day.

F126. *Summary*: **There are a number of ethnic groups in Anatolia that have died out.**

Commentary: The Solymoi were the foes of Bellerophontes (Homer, *Iliad* 6.184, 204) and were thought to be in Lykia or Phrygia (Herodotos 1.173; Strabo 14.3.10). The Leleges were also mentioned by Homer (*Iliad* 10.429, 20.96, 21.86), located in the southwest, and the Bebrykes were a Thracian group that ended up in Mysia (Strabo 12.3.3). The Kolykantioi and Tripsedoi are otherwise unknown.

F127. *Summary*: **The seas around Italy and to the southeast are the Sardoan, Tyrrhenian, Sikelian, and Cretan.**

Commentary: This fragment may belong in a survey of the Mediterranean, as is the case with F104–5. In the *Natural History* it is part of a summary of the various seas within the Mediterranean, for which Eratosthenes is the earliest source mentioned. The Sardoan Sea is defined as everything west of Sardinia, an early attribution before the western Mediterranean was well known. Sardo is the Greek name for Sardinia (Herodotos 1.170), based on an eponymous founder named Sardos (Pausanias 10.17.2): the indigenous name is probably a variant of Shardana, one of the "Peoples of the Sea" of Late Bronze Age Egypt (N. K. Sandars, *The Sea Peoples* [London 1978] 107–8, 113–14, 198–9), as probably were also the Sikels (as Shekelesh, Sandars 112–13). Eratosthenes' Tyrrhenian, Sikelian, and Cretan Seas are more standard nomenclature that remained in use.

F128. *Summary*: **Sailors estimate the distance from Rhodes to Alexandria at about 4,000 stadia or more, but it can be calculated at 3,750.**

Commentary: This is a rare indication that Eratosthenes, who does not seem to have traveled extensively, performed some of his own measurements. The central location of Rhodes and its increasing importance in early Hellenistic times (after it successfully resisted Demetrios Poliorketes in 305 BC: Diodoros 20.81–100) made it a likely place for even a rare traveler to visit. Eratosthenes provided three different distances, good evidence of how he obtained data. There is a sailors' estimate of the direct route, the coastal shipping route, and his own astronomical measurement. A north wind would have increased the distance because of tacking (for ancient tacking, see Lionel Casson, *Ships and Seamanship in the Ancient World* [Princeton 1971] 273–8). A coastal voyage would be about three times the direct distance, depending on whether one went inside or outside of Cyprus.

F129. *Summary*: **It is 2,000 stadia from the Kyrenaia to Krioumetopon, and less from there to the Peloponnesos.**

Commentary: Krioumetopon, the "Ram's Head," is the southwest corner of Crete, modern Cape Krion. The distance from Kyrene is about 320 km. The closest point in the Peloponnesos is Cape Maleia, about 130 km. away, which, as a major sailing point since earliest times (Homer, *Odyssey* 3.287, etc.), is probably the place that Eratosthenes meant. It may be that a distance has dropped out of the text.

F130. *Summary*: **Damastes is incorrect regarding the length of Cyprus. It is a prosperous island with abundant copper mines.**

The island was once overrun with woods and thickets, but mining and government policy have changed this.

Commentary: For Damastes of Sigeion, from the latter fifth century BC, see F13. Eratosthenes objected to his positioning of Hierokepias, which Damastes had evidently placed in the north of Cyprus. Eratosthenes put it in the south, but Strabo, being somewhat picky, pointed out it was in the west. It is actually in the southwest, modern Geroskipou. The most interesting part of the passage is the account of mining on the island, known since prehistoric times. *Chalkanthes* is copper sulfide, widely used medicinally, especially for relief from congestion, ulcers in the mouth, and hemorrhoids. The Cypriot variety was considered the best (Pliny, *Natural History* 34.123–7). The site of Tamassos was important in the Archaic and Classical periods: see Vassos Karageorghis, *Cyprus from the Stone Age to the Romans* (London 1982) 151. Also of interest is the matter of deforestation, although expressed in terms of a virtue rather than an ecological problem. The usual suspects are all there: mining, shipbuilding, and clearing of fields, much like the history of North America. Deforestation had been an issue in the eastern Mediterranean since at least the early Iron Age (Joshua 17.14–18). Eratosthenes' context may be the eighth or seventh century BC, but by Hellenistic times deforestation became a significant matter, although generally seen as a benefit (see Lucretius 5.1370). Yet wood shortages were beginning to develop in many areas (Pliny, *Natural History* 13.95). See further Russell Meiggs, *Trees and Timber in the Ancient Mediterranean World* (Oxford 1982) 371–403, and John F. Healy, *Mining and Metallurgy in the Greek and Roman World* (London 1978) 148–52.

F131–2. *Summaries*: **F131: The distances recorded have been handed down, but cannot be validated, although adjustments have been made as seems necessary. F132: There is little information on western and northern Europe.**

Commentary: Strabo was incredulous that Eratosthenes (and other Hellenistic geographers) knew so little about western and northern Europe, regions that were a vital and evolving part of his own contemporary Roman world. He did admit that one should be tolerant of the fact that Eratosthenes (and the others) had never seen those regions. For Timosthenes see F97: the implication is that his *On Harbors* made some attempt to include the western Mediterranean, since when Strabo wrote that someone was ignorant of a region, his usual meaning was that the source was inaccurate, not lacking in data (see F133). The same may be said about Eratosthenes and the areas cited, although the fragment contains some of the diction of Strabo's day: "Germanika" was probably not used until the first century BC: see F51. Because Strabo would not accept the validity of Pytheas as a source (F14) for Brettanike, he also found Eratosthenes to be ignorant about that region. The other ethnyms and toponyms probably reflect Eratosthenes' own words, and in fact the Bastarnai, a Germanic tribe from the Danube region, were first recorded in his time (Justin 32.3–16). The use of "Keltika" as the conventional term for northwest Europe is also terminology of Eratosthenes' era (F37), received from Pytheas and reflecting an era before Caesar's conquest of Gaul changed the topographical map of the region. Yet despite his objections, Strabo essentially validated Eratosthenes' technique and his use of distances "handed down," rejecting the criticism of Hipparchos. For Dikaiarchos and his geographical treatise see F1.

F133. *Summary*: **It is 300 stadia from Ithaka to Korkyra, 900 from Epidamnos to Thessalonike, and 7,000 from Massalia to the Pillars. The exterior of Iberia as far as Gadeira is populated by the Galatai.**

Commentary: Most of this fragment is a critique of Polybios' objections (34.7) to various data of Eratosthenes, which gives some insight into what Eratosthenes said about Europe. His distances, whether or not they agree with those of Polybios, seem short: Ithaka to Korkyra is perhaps 150 km. (although the sail involves much twisting and turning) and Epidamnos to Thessalonike is about 300 km. (see F65). From Massalia to the current French-Spanish border is a direct distance of about 225 km., and on to the Pillars is about 1,000 km. Land and sailing distances, especially for the latter, would be far longer. It is astonishing that Polybios, with much better data, was still greatly in error. "Those today" would probably be Roman sources of the Augustan period, since Augustus himself lived in Tarragona from 27 to 24 BC, and it was the effective center of the Roman world: one would expect surveys to be made. It is problematic whether Eratosthenes described the Tagos River, as he is not named in the sentence. Strabo has a fuller description (3.3.1), based on Roman information of the latter second century BC: it may have first become known to the Greco-Roman world with the early campaigns of Hannibal (Polybios 3.14.5). Eratosthenes may have been the first to restrict the term "Iberia" to the Iberian Peninsula, as it originally was everything west of the Rhone (Strabo 3.4.9). Pytheas had re-defined it somewhat to mean the western coast of France, as well as the peninsula (F37), and Eratosthenes seems to have reduced it even more (Walbank, *Commentary* vol. 1, pp. 369–70). Eratosthenes described a circuit of Iberia, of which remnants survive in F152–3, based probably on sailing itineraries such as those preserved in the *Ora maritima* of Avienus, written in the fourth century AC. The presumed contradiction in Eratosthenes' description of the Galatai may reflect their diminishing presence in Iberia: by the third century BC they seem to have been limited to north of the Pyrenees, where Polybios consistently located them. On Eratosthenes and the Iberian Peninsula, see José Miguel Alonso-Núñez, "La vision de la Péninsule Ibérique," *Sacris erudiri* 31 (1989–90) 5–8.

F134–6. *Summaries*: **F134: There are three promontories coming down into the Mediterranean from the north: the Peloponnesos, Italy, and the Ligystikan. Timosthenes is a good source for this material, although he has many errors. F135: Europe consists of three promontories, Iberia, Italy, and that down to Maleia. F136. Thrace makes a turn to the south where it touches Makedonia.**

Commentary: As with the western Mediterranean (F133), Eratosthenes had scant knowledge of the southern coast of Europe. Relying on Timosthenes (see F97), Eratosthenes saw southern Europe in terms of peninsulas, a point of view that seems bizarre today but reflects a Greek perspective that saw Italy and Iberia as intruding into the Mediterranean in the same way as Greece. Ligystika (Liguria) was originally the territory around Massalia (Strabo 4.6.3) but seemed to wander both east (Strabo 5.2.1) and west (as Eratosthenes has it). It is a peculiar definition of the Iberian Peninsula not documented elsewhere, and may come from Timosthenes. Most of Strabo's criticism concerns the Greek peninsula, so there are more of Eratosthenes' details preserved in this portion than elsewhere, although there is nothing surviving that is unusual or notable. There seems to have been a certain amount of quibbling as to how many promontories existed and how one was constituted, beginning perhaps with Polybios (34.7.11–14) and continuing with Strabo. If Strabo quoted Eratosthenes correctly in F135, he included ethnic groups far to the north of the actual peninsula, from Greece to the Tanais (Don) River: certainly not a geographical definition but a convenient division for ethnic purposes.

F137. *Summary*: **The Makedonian city of Ichnai is the same as Achne.**

Commentary: For Stephanos' *Ethnika* see F91. Ichnai lies a few kilometers east of Pella.

F138. *Summary*: **The citizens of Euboian Athens are called Athenetai.**

Commentary: Several ancient biographies exist of Aratos of Soloi, the author of the astronomical poem *Phainomena*, of the early third century BC. All of them probably derive from the one by Boethos of Sidon in the second century BC (*Geminos* 17.48; see Marco Fantuzzi, "Aratus" (#4), *BNP* 1 [Leiden 2002] 955). The first biography has a unique citation of Eratosthenes: mention of the obscure city of Athens on Euboia, more properly Athenai Diades, at the northern end of the island, near modern Gialtra and founded by the more famous Athens (Strabo 10.1.5). It was an early member of the Delian Federation (from at least 452 BC: see Benjamin Meritt et al., *The Athenian Tribute Lists* [Cambridge, Mass. 1939] vol. 1, pp. 281–9, 482). The reference is the only surviving detail by Eratosthenes about Central Greece.

F139. *Summary*: **At Helike is evidence of the earthquake two years before Leuktra, where a submerged bronze Poseidon is a danger to fishermen.**

Commentary: Helike, a small town on the Corinthian Gulf in Achaia, was destroyed by an earthquake in the winter of 373 BC. Herakleides of Pontos witnessed the event, and recorded that the city was submerged even though it was 12 stadia from the sea, and that the earthquake was the result of the wrath of Poseidon, because of a dispute over his statue and shrine and perhaps mistreatment of Ionian ambassadors (Strabo 8.7.2; Diodoros 15.49). Earthquakes are common in Greece, to say the least, but this is one of the better-documented ones from antiquity, probably because of Eratosthenes' report. His interest in the formative processes of the earth is apparent throughout Book 1, especially with his reliance on Straton of Lampsakos (F15), and this may have led Eratosthenes to visit the site, the only known field trip connected with his research.

F140. *Summary*: **Near Pheneos the Anias River flows down into sinkholes, which may overflow, but generally the water empties**

into the Ladon and Alpheus, at one time causing flooding around Olympia. The Erasinos around Stymphalos also disappears under the mountains.

Commentary: The usual Greek interest in underground rivers is apparent (see F87, 96). The nature of the Greek landscape means that underground outlets are common, and it was possible to theorize connections of watercourses. The Arkadian city of Pheneos, in the north central Peloponnesos, lies at the edge of a small plain that has no surface outlet. It has alternated as a lake and plain since antiquity: in Eratosthenes' time, unlike today, it was a lake. The underground outlets, locally called "zerethra," a word otherwise unknown, carried the water about 10 km. to the southwest, and down 400 m., to the source of the Ladon, one of the larger rivers of the Peloponnesos, which flows into the Alpheus about 20 km. above Olympia (see James G. Frazer, *Pausanias's Description of Greece* [reprint, New York 1965] vol. 4, pp. 231–3). Eratosthenes' knowledge of an unusual local word suggests a visit (see also F141), and he was at Helike, less than 40 km. away (F139). Just to the east of the plain of Pheneos is Stymphalos, lying in a similar valley that, however, is a permanent lake (although greatly varying in size), famed for its mythological birds. The Erasinos (modern Kephalari) is a small stream of the southwestern Argive Plain, flowing into the sea just north of Lerna, and originating in springs about 15 km. to the west (Pausanias 2.24.6; Diodoros 15.49.5), where there are spectacular caverns (Frazer [supra] vol. 3, pp. 210–11; vol. 5, pp. 564–5). Although the distance from the springs to Stymphalos is about 40 km., they are believed to be connected. The Athenian general Iphikrates was active in the northern Peloponnesos in 391–389 BC, during the Corinthian War (Xenophon, *Hellenika* 4.4.15–16; Diodoros 14.91.2–3). The tale comes from a source unfavorable to him, unlike the other extant material, such as the eulogistic biography by Cornelius Nepos (11.1–4), and may be a local point of view. For some of the textual problems in this passage, see Strabo, *Géographie* 5 (ed. Raoul Baladié, Paris 2003) 211.

F141. *Summary*: **The *phellos* grows in Arkadia.**

Commentary: Although Homer's passage on Arkadia (*Iliad* 2.603–14) does not mention trees, Eustathios used it to mention Eratosthenes' citation of the flora of the region, an area that he seems to have described in detail (F139–40). The trees are all oaks: the *prinos* (Hesiod, *Works and Days* 436) and *drys* (Homer, *Iliad* 11.494; *Odyssey* 14.328) had been identified in early times, the latter the one associated with Zeus of Dodona. The *phellos* was not mentioned until later (Theophrastos, *Research on Plants* 1.5.2). *Thelyprinos* ("woman's prinos") is a unique Arkadian localism, perhaps indicating Eratosthenes was personally familiar with the region (see also F140).

F142–4. *Summaries*: **F142: The Agraeis are mentioned. F143: The Taulantioi live near Dyrrachion, also called Epidamnus. The graves of Kadmos and Harmonia are nearby. F144: The Autariatai are a Thesprotikan people.**

Commentary: For Stephanos of Byzantion and his *Ethnika* see F91. These three entries all concern the ethnic makeup of Illyria. The Agraeis, or Agraioi, were an independent monarchy in the fifth century BC and minor players in the Peloponnesian War (Thoukydides 3.111), living in the country east of the Ambrakian Gulf. Epidamnos had been founded in the seventh century BC by the Kerykians and Corinthians: disputes over it was one of the main causes of the Peloponnesian War (Thoukydides 1.23–6). Around 300 BC it came to be called Dyrrhachion, the name of a topographical feature overlooking the city. In 229 BC it came under Roman control (Polybios 2.11), but there is no evidence that Eratosthenes was aware of this. The Taulantioi had long been known as the primary indigenous group of the region (Hekataios [*FGrHist* #1], F99, 101). The Drilon (modern Drin) and Aoos (modern Vjosa) are at the north and south extremities of the long narrow plain in which Epidamnos lay. By Hellenistic times the graves of Kadmos and Harmonia were believed to be in the area (Apollonios of Rhodes 4.517). The Autariatai lived farther north, in what is now Bosnia, and were the most powerful regional group (Strabo 7.5.11).

F145. *Summary*: **The Nestaioi are among the Illyrians.**

Commentary: The Nestaioi are first mentioned in the fourth century BC (Pseudo-Skylax 23), and then by Apollonios of Rhodes (4.337, 1215). They lived on the Illyrian mainland near the island of Pharos (modern Hvar in Croatia), an outpost of Paros (Diodoros 15.13.4).

F146. *Summary*: **The Hyllis peninsula is as large as the Peloponnesos and is populated by Greeks who have taken up barbarian ways.**

Commentary: The Hyllis peninsula is the promontory (modern Punta Planka) west of Split in modern Croatia, by no means close to the size of the Peloponnesos. The story is a typical Heraklid foundation myth, attributing people of obvious indigenous characteristics to a Greek origin and explaining it by local assimilation. The Hylloi were first mentioned in the fourth century BC (Pseudo-Skylax 22). Although Eratosthenes is specifically cited and was a source that the author of the *Periplous* respected (line 114), the ethnographic detail is more in the style of Timaios (*FGrHist* #566) and his interests in foundation myth. Eratosthenes' contribution may have been a brief recognition of the place and its people, as in F145 and 147.

F147. *Summary*: **The Teriskoi are around Mt. Alpeion.**

Commentary: The Teriskoi (or Tauriskoi) lived in what is now northeastern Italy and into Austria and were known for their gold mines (Polybios 34.10.10). The Alps were still only vaguely known in Eratosthenes' time: it was only at the end of the third century BC that they became better understood, although Eratosthenes represents some advance over Herodotos (4.49), who knew them only as a river.

F148–9. *Summaries*: **F148: The two mouths of the Istros empty into the sea around the island of Peuke. F149: The Istros flows from uninhabited territory and around the island of Peuke.**

Commentary: The Istros (Danube) emptied into the Black Sea in a vast delta that, along with the Indos and Nile, was one of the great river mouths of classical antiquity. There were at least six known mouths. The region around Peuke ("Pine") Island had become well known to Greeks ever since Dareios of Persia had constructed his Istros bridge there in the late sixth century BC, made famous by Herodotos (4.83–93) but also commemorated in an epic poem of the late fifth century BC, the *Crossing of the Pontoon Bridge* by Choirilos of Samos. Just after becoming king, Alexander the Great made an expedition to the island (Strabo 7.3.8). The comparison with Rhodes is baffling, unless it merely refers to the pine trees. Attempts to locate Peuke are frustrated by changes in topography: in Strabo's day (or that of his unnamed source) it was 120 stadia inland up the southernmost mouth of the river (7.3.15).

F150. *Summary*: **The Orkynian woods are the most fertile part of Germany.**

Commentary: Julius Caesar is the earliest extant source to have quoted the *Geographika*. He equated the Hercynian woods with Eratosthenes' Orkynian, which, assuming the variants in nomenclature refer to the same toponym, are probably the "Arkynian mountains" of Aristotle (*Meteorologika* 1.13), located among the northward flowing rivers of Europe, and thus well beyond the Danube. Caesar's description (*Gallic War* 6.25–8) places it along the river, a vast forest more than nine days across. It probably refers to the hilly and forested country north of the Alps in Austria and the Czech Republic.

F151. *Summary*: **Neither Kyrnos nor Sardon can be seen from the mainland.**

Commentary: Strabo described how from a vantage point at Popolonion in Italy (Roman Populonium, modern Piombino), he had seen Sardo (Sardinia) far away, Kyrnos (Corsica) nearer, and closer still Aithalia (Elba). Thus he disputed Eratosthenes' report that the two larger islands were not visible from the Italian coast. But it is unlikely that Eratosthenes' informant was that far north, and more probably in the region of the Bay of Naples, where it is unlikely that any of the islands could be seen.

F152. *Summary*: **Tarrakon has an anchorage.**

Commentary: Most of Strabo's comments describe the situation in the first century BC: the memorial that Pompeius set up in the Pyrenees after his campaign against Sertorius in the 70s BC (Cassius Dio 41.24.3), and the suitability of Tarrakon (Roman Tarraco, modern Tarragona) as a residence for commanders, probably referring to Augustus' stay in the city from 27 to 24 BC. Only the comment on Ebysos (modern Ibiza in the Balearic Islands) can be attributed to Eratosthenes: the island was still under Carthaginian control in his day.

F153. *Summary*: **The territory adjoining Kalpe is Tartessis, and Erytheia is the Island of the Blessed. It is five days from Gadeira to the Sacred Promontory, where the tides terminate. Northern Iberia is a better route to the Keltic territory than by sea. All this was reported by Pytheas.**

Commentary: The passage records an itinerary from Kalpe (the northern Pillar of Herakles, modern Gibraltar: see F106) along the Iberian coast to the Keltic territory (western France). Because the ultimate source is Pytheas of Massalia, whose veracity Strabo always rejected (see F14), the account is confused and has suffered from Strabo's editing. Tartessis (or Tartessos) was the wealthy southwestern coast of Spain, perhaps biblical Tarshish (Genesis 10.5) which produced tin and silver and became known to the Greek world in the latter seventh century BC (Herodotos 4.152), providing a great source of wealth to Archaic Greece. The major city of the region was the Phoenician trading post at Gadeira (modern Cádiz). Beyond was the Sacred Promontory (perhaps modern Cabo de San Vicente, the southwest corner of the Iberian Peninsula), but the information on the tides is hopelessly confused, as they were well known in this region and actually increase as one goes north. Strabo has misunderstood or misrepresented data about tides farther north, about which Pytheas was the earliest source of information. There is also a hint of a land route to Keltika, which Pytheas used, following an ancient trade path from Massalia up the Rhone and down the Loire. In addition, Polybios (3.57.3) cited unnamed authors who discussed the Pillars, the External Ocean, and the mines of the Iberian Peninsula and Britain, which may be vestiges of Eratosthenes' fuller treatment of Iberia: see Walbank, *Commentary* vol. 1, p. 394. Despite Strabo's editing, F153 preserves some significant facts that Eratosthenes obtained from Pytheas' *On the Ocean* (see further, Roller, *Pillars* 3–7, 69, 76–7).

F154. *Summary*: **Banishment of foreigners is a common feature of all barbarians, but the Egyptians are especially censured for this, even using Homeric evidence. Yet the Karchedonians and Persians had the same traits.**

Commentary: To conclude the *Geographika* Eratosthenes addressed questions of ethnicity and the tendency of one group to exclude all others. The first example is from Egypt, where mythical King Bousiris would sacrifice all foreigners (for the scattered sources for this tale, see Ruth Ilsley Hicks, "Egyptian Elements in Greek Mythology," *TAPA* 93 [1962] 102–8), a story given a vague Homeric authority and connected with the shepherd-pirates who would come to inhabit Greek novels. It was common knowledge that the Carthaginians took extreme measures to keep shipping from territory that they believed was their own: a treaty with Rome from probably the late sixth century BC insisted on this

(Polybios 3.22). This limited Greek exploration in the western Mediterranean and beyond, probably resulting in some inaccurate information, although Strabo may have been making one of his regular swipes at Pytheas (see F14). Whether the comment about the Persians comes from Eratosthenes is not certain. The first two examples of inter-ethnic hostility provide a contrast with the ecumenical attitude expressed in F155, the end of the *Geographika*.

F155. *Summary*: **It is better to distinguish people by their good and bad characteristics, rather than whether they are Greek or barbarian. Alexander was urged to make the latter distinction, but refused to do so.**

Commentary: Although the examples of F154 are extreme, Greek history is replete with instances of the superiority of one's own ethnic group over all others. The very prejudice is contained in the word "barbarian," meaning "someone who cannot speak properly," a word that came into common use in the fifth century BC (Aischylos, *Persians* 255; Herodotos 1.58; Homer used the more cumbersome but more explicit βαρβαρόφωνος, *Iliad* 2.867) Yet the new world of Alexander and the Successors, in which Greeks were only a tiny part of the known world, and which revealed sophisticated cultures that knew nothing of Greece, made such views less viable: F155 notes several non-Greek groups that were nonetheless capable and virtuous. Although this passage has long been interpreted to mean that Eratosthenes spoke highly about the Romans (see, for example, T. J. Cornell, "The Conquest of Italy," *CAH* vol. 7, part 2 [second edition, 1989] 419), this is unlikely, given his general disinterest in Rome. The reference to their political expertise sounds like Strabo's gloss from the viewpoint of the Augustan period. On the other hand, the general excellence of the Carthaginians and their institutions had been noted since the fourth century BC (Isokrates, *To Nikokles* 24; Aristotle, *Politics* 2.8). Alexander, however, had been advised to adhere to the conventional view and treat barbarians as if they were plants or animals, holding strongly to Greek superiority. Yet he chose not to do so, even suggesting that national styles of dress be abolished and believing that the entire inhabited world was the country of everyone, a complete reversal of the meaning of οἰκουμένη, which originally applied to only the Greek world (Demosthenes, *On Halonnesos* 35; *On the Crown* 48). These views were set down by Zenon of Kition in his *Republic*, who saw humanity as a world community sharing the nourishment of a common field (Plutarch, *On the Fortune and Virtue of Alexander* 6–8). Zenon's student Eratosthenes incorporated the ideas in his *Alexander* (a source for Plutarch's essay), and then used the theme as a fitting end to his *Geographika*, which probably discussed more ethnic groups than had any previous treatise. On difficulties with this passage see A. B. Bosworth, "Alexander and the Iranians," *JHS* 100 (1980) 3–4; also W. W. Tarn, *Alexander the Great* 2: *Sources and Studies* (Cambridge 1948) 437–49.

Gazetteer

Approximately 400 toponyms are preserved in the *Geographika*. They range from the obvious (Athens, Europe) to the obscure and unlocated. The following gazetteer has been designed to assist with identification and to limit intrusive toponymic details in the commentaries.

Identification of ancient toponyms is a problematic and often messy business. Even the ones that may seem obvious (e.g. Albania, India) may bear little or no relationship to their modern homonyms. The complex recension of the fragments applies equally to their toponyms and only exacerbates an already difficult problem. Spellings are inconsistent, often reflecting the original source, yet there may be similar or even identical names for widely separated places, a problem particularly apparent with Hellenistic royal toponyms, some of which can number in the dozens. Ethnyms may be used topographically. The limitations of ancient toponymic research may result in interpretations more misleading than informative. Since antiquity it has been standard to reinterpret toponyms within the nomenclature of one's own culture: this first became conspicuously apparent when the Romans began to write about the Greeks. Yet their use of Latin for Greek names was inconsistent and can be baffling. In early modern times this desire to reinterpret became rampant, with many editors, especially in the nineteenth and early twentieth centuries, substituting, often erroneously, modern toponyms for ancient ones. This trend also had its political aspects, especially in Greece and Italy, as medieval and early modern names were replaced with presumed ancient ones, with an astonishingly high degree of inaccuracy. Thus modern place-names that appear to be survivals from antiquity are often nothing of the sort, or, commonly, in the wrong location. The continued obsession of contemporary classicists with latinization of proper names, even creating Latin forms that did not exist in antiquity, only makes a complex situation worse.

The following gazetteer attempts to locate all the toponyms mentioned in the text, although this is impossible in some cases. Where appropriate (e.g. Rome) it describes how an especially familiar name fit into Eratosthenes' scholarly world. Toponymic history has been kept to a minimum and, if significant, is discussed in the appropriate commentaries. Ethnyms generally do not appear in the gazetteer unless there is a strong topographic component. Not every toponym listed may have been used by Eratosthenes, and this is indicated wherever possible. All toponyms in this gazetteer that may have appeared in the *Geographika* are in all capitals. Many of the toponyms appear in *BA*, but a number of them are in areas not covered by the *Atlas*, or cannot be located with certainty.

ABILYX (F106, Map 3), the southern PILLAR OF HERAKLES, more commonly Abila, modern Jebel Mousa in Morocco.

ACHEROUSIAN MARSH (F7, Map 3), a tidal estuary in the vicinity of Baiai, just west of Naples, modern Lago di Fusano.

ACHNE (F137, Map 4), Makedonian city, allegedly the same as ICHNAI, 4 km. east of Pella.

ADRIA, ADRIAS, ADRIATIC GULF (F13, 16, 131, 134, 135, Maps 3, 4), the Adriatic.

AEGEAN (F11, Map 4), the sea east of the Greek peninsula.

AFRICA (F105), the Roman name for the continent of LIBYA and, more specifically, its north central portion. It is unlikely that Eratosthenes used this term.

AIGINA (F16, Map 4), island in the Saronic Gulf, 30 km. southwest of Athens. In Eratosthenes' day it was controlled by various of the major powers.

AILANA (F95, Maps 5, 6), city at the head of the AILANTIC GULF, a trading emporium for the commodities of ARABIA, at the site of modern Aqaba in Jordan.

AIOLOS (F7, Map 3), actually a personal name, one of the opponents of Odysseus (Homer, *Odyssey* 10.1–79), but used topographically, referring probably to the Aiolian (modern Lipari) Islands north of Sicily.

AITHIOPIA (F8, 20, 37, 40, 43, 74, 92, 95, 98, 100, 107, Maps 2, 3, 4), originally the territory south of EGYPT, used generically by Homer (*Odyssey* 1.23) for the farthest known peoples. Its capital was MEROË. Eventually the term was expanded to include all the known parts of sub-Saharan Africa south and west of EGYPT (but not including the territory of modern Ethiopia): the peoples of West Africa were called Western Aithiopians (F107). The AITHIOPIAN SEA (F38) is the western Indian Ocean and was probably not a term used by Eratosthenes.

AITNA (F6, Map 3), volcano in eastern Sicily, modern Mt. Etna.

AKAKESION (F8, Map 4), topographical feature and city in ARKADIA, west of Megalopolis, associated with Hermes.

AKAMAS (F130, Map 5), the northwestern promontory of Cyprus, modern Cape Arnaouti.

AKARNANIA (F142, Map 4), the territory of coastal west central Greece, from the Gulf of Corinth to the Ambrakian Gulf. In Eratosthenes' day its cities were a federal league.

ALBANIA (F109–11, Map 5), coastal district on the west side of the Caspian Sea, northwest of modern Baku in Azerbaijan, the beginning of a trade route to the Black Sea, not to be confused with the modern country on the Adriatic.

ALEXANDRIA AMONG THE AREIOI (F78, 108, Map 7), city founded by Alexander near modern Herat in Afghanistan.

ALEXANDRIA BY (or IN) EGYPT (F25, 34–5, 41, 56, 59–60, 65, 67, 100–2, 128, Maps 1, 2, 5), city founded by Alexander in 331 BC, which has retained its name and importance until today. As Eratosthenes' home, one of the places vital to his measurement of the earth, and where his main meridian crossed a parallel, it was crucial to his research.

ALEXANDRIA TROAS (F60, Map 4), the name given around 300 BC to Antigoneia, lying on the west coast of the TROAD, modern Dalyanköy. In Eratosthenes' day it was a Pergamene possession and was on one of his parallels.

ALPES, ALPEION (F51, 147, Map 3), a generic term for the mountains of EUROPE, known to the Greek world since the fifth century BC (Herodotos 4.49). In Eratosthenes' day they were still vague and undefined, including everything from the Pyrenees to the northern Balkans.

ALPHEUS (F140, Map 4), the major river of the western PELOPONNESOS, flowing past OLYMPIA.

AMAXA (F123, Map 5), a district of BITHYNIA in Anatolia.

AMISOS (F47, 50–1, 61, Map 5), Greek city on the Black Sea in Anatolia (modern Samsun), lying on a meridian. In Eratosthenes' time it became a royal city of the emergent Pontic kingdom.

AMMON (F15–16, Map 5), oracle at the oasis of Siwa in western EGYPT, famous to Greeks since early times but especially so after Alexander's visit of 331 BC.

AMPHIPOLIS (F60, Map 5), Athenian foundation in northern Greece, lying on a parallel, and an important Makedonian possession in Eratosthenes' day.

ANIAS (F140, Map 5), river at PHENEOS in the northern PELOPONNESOS, presumably one of the small streams with no outlet to the sea in the plain west of Mt. Kyllene.

ANTHEDON (F2, Map 5), Boiotian city mentioned in the Homeric *Catalogue of Ships* (*Iliad* 2.508), modern Loukisia.

AOOS (F143, Map 5), major river of Illyria, the modern Vjosa in modern Albania.

AORNOS (F7, Map 3), Greek form of Avernus, a volcanic crater with a lake, modern Lago Averno just north of the Bay of Naples.

APOLLONIA (F60, Maps 3, 4), city in Epeiros, near Pojani in modern Albania. In Eratosthenes day' it became one of the first mainland Greek cities to come under Roman control.

ARABIA (F8, 81, 92–8, Maps 1, 2, 5, 6), essentially the territory of the modern Arabian peninsula, but also including the wedge of desert territory from east of the Jordan River to MESOPOTAMIA, as far north as SYRIA. In Eratosthenes' time it was a number of indigenous states, with PETRA and the Nabataeans in the north and the wealthy aromatic-producing kingdoms of ARABIA EUDAIMON ("Happy Arabia") in the south, all of which traded extensively with the Mediterranean world.

ARABIAN GULF (F10, 13, 57, 59, 92, 95, 97, Map 6), the ancient name for the Red Sea.

ARACHOSIA (F71, 78, Map 7), district now in modern Afghanistan and eastern Iran, conquered by Alexander in 329 BC and then a Seleukid satrapy until transferred to Indian control in Eratosthenes' time. Its capital was Alexandria in Arachosia, near modern Kandahar.

ARADOS (F94, Map 6), island in the PERSIAN GULF, modern Muharreq, one of the islands of Bahrain.

ARBELA (F83, Map 5), city in eastern Assyria at the edge of the Iranian plateau (modern Erbil in Iraq), on the main route east. Alexander was there in 331 BC, and in Eratosthenes' day it was under Seleukid control.

ARBIS RIVER (F77, Map 7), a stream west of the INDOS, perhaps the modern Hab, at Karachi in Pakistan.

ARGIVE TERRITORY (F33, 140, Map 4), the area of the Greek city of Argos in the PELOPONNESOS.

ARIA, ARIANA (F63, 66, 69, 77–80, 82, Maps 1, 2, 5), as defined by Eratosthenes the territory between INDIA and MESOPOTAMIA, essentially the Iranian plateau, a vast region of many ethnic groups. In his day the eastern part was under Indian control and the western was Selukid. Its eponymous city, ARIA, was probably at ALEXANDRIA OF THE AREIOI, and the toponym ARIA/ARIANA originally referred only to this district, around modern Herat in Afghanistan.

ARKADIA (F141, Map 4), a district of the central Peloponnesos.

ARMENIA (F13, 15, 48, 63, 83, 87, 120, Map 5), a district of eastern Anatolia, a remote and mountainous territory still little known in

Eratosthenes' day, ruled by an indigenous dynasty vassal to the Seleukids.

ARMENIAN GATES (F83, Map 5), location uncertain, but one of the means of access from upper MESOPOTAMIA into ARMENIA, probably along the EUPHRATES.

ARMENIAN MOUNTAINS (F63), a generic term for the mountains of ARMENIA.

ASIA (F16, 24, 33, 47, 53, 71, 95, 100, 109, 126, Map 2), the largest of the three continents, extending from Anatolia to INDIA and the Eastern Ocean, and north to the Caspian and Black Sea and points beyond.

ASTABORAS, ASTAPOUS, ASTASOBAS (F98, Map 6) various affluents of the NILE. Since the ASTABORAS is below MEROË, it is probably the modern Atbara in Sudan. The ASTAPOUS or ASTASOBAS has the characteristics of the White Nile.

ATHENS (F8, 33, 47, 50, 54, 56, 63, 83, Map 4), famous Greek city, lying on Eratosthenes' main parallel. There was another ATHENS (F138), an obscure town in northern EUBOIA.

ATLANTIC OCEAN (F30, 33, 39, 69, 95, Map 3), the western EXTERNAL OCEAN, accessed through the PILLARS OF HERAKLES.

ATLAS MOUNTAINS (F100, Map 3), the mountains of northwest AFRICA, extending in an arc from the western Mediterranean to the ATLANTIC, dominated by Mt. Atlas (modern Jebel Toubkhal in Morocco), at 4,167 m. the highest mountain in northern AFRICA.

ATTIKA (F11, 47, 138, Map 4), the territory around ATHENS.

AULIS (F8, Map 4), eastern Boiotian sanctuary of Artemis, famous in the events of the Trojan War.

AUTOMALA (F104, Maps 3, 4), or Automalax, fortress at the southernmost point of the Mediterranean and the western boundary of the KYRENAIKA.

AXEINOS, AXINOS (F8, Maps 4, 5), primitive name for the Black Sea, believed to mean "inhospitable" but more probably derived from a Persian or indigenous term.

AZANIAN SEA (F93, Map 6), term for the western reaches of the Indian Ocean. AZANIA itself was in the region of modern Somalia, and the name may survive in modern Zanzibar.

BABYLON, BABYLONIA (F39, 60, 62–4, 80, 83–5, 87, 89, 90, 94–5, Map 5), the ancient Mesopotamian city and its territory; in Eratosthenes' day a Seleukid center.

BAKTRIA or BAKTRIANA (F47, 50, 53, 60, 78, 81, 108, 131, Map 7), a vast region northwest of INDIA on the upper OXOS (modern Amu

Darya), centered at the city of BAKTRA (modern Balkh in Afghanistan). In Eratosthenes' day the Seleukid satraps were beginning to form the Greco-Baktrian kingdom.

BERENIKE (F41–2, 59, Map 6), important trading port on the Red Sea for the voyage to INDIA, founded by Ptolemaios II, at modern Medinat el-Haras.

BITHYNIA (F123, Map 5), Greek kingdom in northwest Anatolia.

BLESSED ISLANDS (F153), mythical locale where the chosen dead resided, always just beyond the limits of known geography. In Eratosthenes' day they were in the ATLANTIC and equated by him with ERYTHEIA at GADEIRA.

BORYSTHENES (F8, 25, 34–6, 61, Map 5), major river (the modern Dneiper) flowing into the Black Sea from the north. At its mouth were a number of Greek settlements often collectively called Borysthenes, which lay on Eratosthenes' main meridian and one of his parallels.

BOSPOROS (F61, 115, Maps 4, 5), the passageway leading into the Black Sea from the PROPONTIS, past BYZANTION.

BOUSIRIS (F154, Map 5), city and district in the central Egyptian Delta, on the mouth of the NILE of the same name.

BRETTANIKE (F14, 34–5, 131, Map 3), Greek name for Great Britain, probably an error for the original PRETTANIKE.

BYZANTION (F15–16, 51, 60–1, Maps 4, 5), Greek city on the BOSPOROS (modern Istanbul), lying on Eratosthenes' main meridian.

CARTHAGE: see KARCHEDON.

CHALKIDEAN STRAIT (F16, Map 4), the narrow sea passageway at Chalkis in central Greece, known as the Euripos.

CHALONITIS (F89, Map 5), the area south of Lake Van in Turkey, near the course of the TIGRIS.

CHARYBDIS: see SKYLLA.

CHATRAMOTITIS (F95, Map 6), an aromatic-producing district of southern ARABIA, perhaps the modern region of Hadhramaut in Yemen.

CHERSONESOS (F143), referring to the area around EPIDAMNOS (DYRRHACHION), Durres in modern Albania, a term probably not used by Eratosthenes.

CHOASPES (F13, Map 5), the river that flows by SOUSA, one of the tributaries of the modern Karun in Iran.

CRETE (F127, Map 4), the island at the south end of the Aegean.

CRETAN SEA (F15, 127, Map 4), the sea between CRETE and the southern Aegean islands.

CYPRUS (F130, Map 5), the island in the northeastern Mediterranean, in Eratosthenes' day under Ptolemaic control.

DEIRE (F95, Map 6), promontory and village on the African side of the entrance to the Red Sea, where the straits are narrowest, probably modern Ras Siyan at the Bab el-Mandeb.

DEMOS (F8), presumed town on ITHAKA, but probably a misunderstanding of Homer, *Iliad* 3.201.

DIKAIARCHIA (F7, Maps 3, 4), Greek settlement on the north side of the Bay of Naples, later Roman Puteoli, modern Pozzuoli.

DIOSKOURIAS (F13, 52, Map 5), Greek city on the eastern shore of the Black Sea and an important trading port for the Caucasus, modern Sukhumi in Georgia.

DRANGIANE (F78, Map 7), the territory of the Drangai, an indigenous population of the eastern Persian Empire, and later under Seleukid control, the area of the modern Helmand basin on the Afghan-Iran frontier.

DRILON (F143, Map 4), river of ILLYRIA, probably the modern Drin in northern modern Albania.

DYRIS (F100, Map 3), local name for MT. ATLAS, a corruption of the Berber "Idrarn."

DYRRHACHION (F143, Maps 3, 4), topographical feature of the coast of ILLYRIA, which became the name of the town of EPIDAMNOS in early Hellenistic times, Durres in modern Albania. In Eratosthenes' day it was an early outpost of Roman control in Greece.

EBYLOS or EBYSOS (F152, Map 3), island in the western Mediterranean, a Carthaginian outpost, modern Ibiza.

EGYPT (F2, 8, 15–16, 33, 35, 55, 60, 62, 74, 92, 95, 98, 100, 154), the ancient land along the NILE, from the Delta at the Mediterranean south to SYENE at the First Cataract. Although territorial boundaries were fluid, it exercised control over territory both well east and west of the river.

EGYPTIAN ISLAND (F35, 53, 98, Map 6), more properly "Island of the Fugitive Egyptians" (F53), held by a group of Egyptian soldiers who revolted from King Psammitichos (probably II, 595–589 BC); the story is told by Herodotos (2.30–1), who placed the island two months upriver from MEROË, which would be near or above Khartoum.

EGYPTIAN SEA (F8, 10, Map 5), refers to that part of the Mediterranean off the Egyptian Delta and the Sinai Peninsula, used by Eratosthenes only in a Homeric context.

EKBATANA (F83, Map 5), the ancient Median capital, modern Hamadan in Iran, a Seleukid center in Eratosthenes' day.

EKREGMA (F33, Map 5), the "Outlet," perhaps more descriptive than toponymic, referring to the outlet from LAKE SIRBONIS to the sea, in eastern lower EGYPT.

EMODOS (F69–70, Map 6) or EMODON MOUNTAINS, high range north of INDIA: the description of Ammianus Marcellinus (23.6.64) shows that in his time, at least, the highest Himalayas were meant.

EMPORIKOS (F100, Map 3), gulf on the West African coast, originally the location of a Carthaginian trading post, but under Greek control in Eratosthenes' day, probably the estuary of the modern Oued Sebou in Morocco.

EPIDAMNOS (F65, 133, 143, Maps 3, 4), Greek city on the Illyrian coast, officially DYRRHACHION in Eratosthenes' time, Durres in modern Albania.

ERASINOS (F140, Map 4), river in the western Argive plain in Greece, assumed by Eratosthenes to have its origin at STYMPHALOS.

ERYTHEIA (F153, Map 3), the island on which GADES/GADEIRA (modern Cádiz in Spain) is situated, which Eratosthenes thought was the location of the BLESSED ISLANDS.

ERYTHRAIAN SEA (F8, 15–17, 33, 83, 93–5, 98, Map 6), the normal Greek term for the western part of the Indian Ocean, often including its two main northern extensions, the PERSIAN GULF and the modern Red Sea (sometimes called the ERYTHRAIAN GULF, but more normally the ARABIAN GULF).

ETEONOS (F8), Boiotian town mentioned in the Homeric *Catalogue of Ships* (*Iliad* 2.497), location unknown.

EUBOIA (F138, Map 4), island of central Greece.

EUPHRATES (F13, 37–8, 62–3, 71, 82–4, 87–90, 94, Map 5), the largest river of Mesopotamia.

EUROPE (F14–16, 24, 33, 37, 100, 132–5, Maps 1, 2), one of the three continents, the territory lying north of the Mediterranean.

EUXEINOS (F8, 94, 109, 118, 135, Maps 1, 4, 5), the Black Sea. The term was believed to mean "hospitable" but probably was from Persian *aesaena*, "dark" (i.e. "black"). PONTOS, specifically the territory along the south shore of the sea, was also a name for the sea itself.

EXTERNAL SEA (F9, 15–17, Map 3), the ancient term for the OCEAN beyond the PILLARS OF HERAKLES. By Eratosthenes' day it was somewhat anachronistic, as localized names (ATLANTIC, ERYTHRAIAN) had come into use.

GADEIRA, GADES (F14, 38, 53, 106, 133, 153, Maps 1, 3), ancient Phoenician trading post on the southwest coast of Spain (modern Cádiz), under Carthaginian control in Eratosthenes' day yet an important

Greek point of access to the world beyond the PILLARS OF HERAK-
LES. GADEIRA is the Greek form (Phoenician Gadir); GADES is the
Roman version and was probably not used by Eratosthenes.

GADEIRAN GATES (F106, Map 3), early term for the Straits of Gibraltar, probably obsolete in Eratosthenes' day.

GALATA (F105, Map 3), small island off the north African coast,
probably modern Galite north of Tunisia.

GANGES (F38, 69, 74, Maps 1, 7), the largest river of INDIA, whose
lower course was little known but whose mouth was the easternmost
point in Eratosthenes' scheme of the inhabited world.

GANGRA (F122, Map 5), capital of Paphlagonia, modern Çankırı in
Turkey.

GAUDOS (F8, 9, Maps, 3, 4), island in the Mediterranean, thought
by some to be Kalypso's island. Its location is uncertain: perhaps modern
Kavlos south of CRETE or, more probably, GAULOS (modern Gozo).

GAUGAMELA (F83, Map 5), village in northeastern Assyria, famous
because of Alexander's defeat of Dareios III nearby in 331 BC, and lying
on a main route between Mesopotamia and Media, perhaps modern Tell
Gomel in Iraq.

GAULOS (F105, Map 3), island in the Mediterranean, modern Gozo,
part of Malta. In Eratosthenes' time it became a Roman possession.

GAZA (F95, Map 5), wealthy and important trading emporium of
the coastal southern Levant, where commodities from ARABIA and interior ASIA reached the Mediterranean. In Eratosthenes' day it was a
Ptolemaic possession; it retains its ancient name.

GEDROSIA (F59–60, 78, 95, Map 7), territory between Persia and
INDIA, near the coast, essentially modern Baluchistan, in Pakistan and
Afghanistan.

GERENA (F8, Map 4), uncertain place, perhaps the Gereneia of
Pausanias (3.26.8) on the southern coast of the Peloponnesos.

GERMANIKA (F131), vague district of north central Europe, a term
not yet in use in Eratosthenes' day. The same can be said for the GER-
MANIKAN MOUNTAINS (F51).

GERRHA (F15–16, Map 5), village on the Mediterranean just east
of the PELOUSIAC MOUTH of the NILE, perhaps modern Tell
Mahmudiyeh.

GERRHA (F94, Map 6), trading center on the west coast of the
PERSIAN GULF, location uncertain, perhaps modern Thaj in Saudi
Arabia.

GETAI (F8, 131, Map 4), extensive ethnic group whose component
parts extended from east of the Caspian (e.g. the Massagetai) to central

Europe. To Eratosthenes, the "territory of the Getai" was the lower Danube and the Balkans.

GLAUKOPION (F8, Map 4), topographical feature of ATHENS, probably one of the precipitous points of the Akropolis.

GORDYAIA or GORDYAIENE (F83, 87, 89, Map 5), remote and mountainous region north of MESOPOTAMIA and south of modern Lake Van, little known in Eratosthenes' day.

GYMNESIAN ISLANDS (F152, Map 3), Carthaginian islands in the western Mediterranean, the modern Balearics.

HALIARTOS (F2, 8, Map 4), Boiotian town mentioned in the Homeric *Catalogue of Ships* (*Iliad* 2.503), modern Kastri Maziou.

HALYS (F8, Map 5), major river of central Anatolia, flowing into the Black Sea, the modern Kızılırmak.

HARMOZAI (F94, Map 6), village on the eastern side of the PERSIAN GULF at its entrance and narrowest point; the name survives in the modern Straits of Hormuz.

HEKATOMPYLOS (F108, Map 5), city southeast of the Caspian Sea, near the location of Alexander's final defeat of Dareios III in the summer of 330 BC. Its location is uncertain but it may be at modern Shahr-i-Qumis in Iran.

HELIKE (F139, Map 4), city in Achaia on the Corinthian Gulf, destroyed by an earthquake in 373 BC and visited by Eratosthenes, near modern Aigion.

HELLAS (F2–3, 8, 33, Map 4), general name for Greece, used topographically to mean the Greek peninsula.

HELLESPONT (F15, 35–6, 47, 50, Map 4), the strait from the upper AEGEAN into the PROPONTIS and leading eventually to the Black Sea.

HERAKLEOTIC MOUTH (F56, Map 5), Greek name for the western mouth of the NILE, more usually called the KANOBIC.

HERA'S ISLAND (F106, Map 3), small islet at the PILLARS OF HERAKLES, location uncertain, either Palomas near Algeciras in Spain or Peregil off the Moroccan coast.

HEROONPOLIS (F56, 95, Map 5), city on the PELOUSIAC MOUTH of the NILE, upstream from PELOUSION, but F47 must refer to the HEROONPOLIS GULF, the modern Gulf of Suez.

HESAIA (F109, Map 7), region of central ASIA.

HESPERIDES (F8, 104, Map 5), mythical place in the far west where the daughters of Hesperis guarded the tree with the golden apples, but also in Eratosthenes' day a lush region in the western KYRENAIKA near modern Coefia in Libya.

HIEROKEPIAS (F130, Map 5), or Hierokepis, village in southwest CYPRUS, modern Geroskipou.

HIERON (F53, Map 3), promontory on the southwest coast of Portugal, probably modern Cabo de San Vicente, but the name may indicate some vague knowledge of IERNE (Ireland).

HYPANIS (F8, Map 5), river flowing into the Black Sea from the north, the modern Bug.

HYPASIS (F41, Map 7) or Hyphasis, river of INDIA and affluent of the INDOS, the modern Beas.

HYLLIS (F146, Map 3), Illyrian peninsula, the modern Punta Planka west of Split in Croatia.

HYRKANIA (F50, 60, 109, 131, Map 5), coastal region at the southeast corner of the Caspian Sea, under Seleukid control, along the modern Gurgan River in Iran.

HYRKANIAN SEA (F24, 47, 53, 109–10, 131, Map 5), the Caspian Sea. The term HYRKANIAN, perhaps originally an ethnym, seems interchangeable with KASPIAN.

IAXARTES (F24, 108, 110–11, Maps 1, 7), river of central Asia, the modern Syr Darya, flowing into the Aral Sea.

IBERIA, IBERIKA (F6, 11, 14, 33–4, 37, 39, 53, 131, 133, 135, 153, Map 3), the Iberian Peninsula, but the name can mean the entire west of EUROPE beyond the Rhone. IBERIKA only occurs at F37, 131, and 133, and may reflect a particular source.

IBEROS (F152, Map 3), or Hiberos, river of northern Spain, the modern Ebro.

ICHNAI (F137, Map 4), city in MAKEDONIA just east of Pella, perhaps modern Kouphalia.

IDA (F6, Map 4), mountain in the TROAD, modern Kaz Dağ.

IERNE (F37, 53, Maps 1, 2, 3), toponym in the far northwest of the inhabited world and on a northern parallel, clearly Ireland, but only vaguely known.

IKAROS (F94, Map 5), island in the upper Persian Gulf, modern Failaka in Kuwait.

ILION (F6, Map 4), a name for the site of TROY, assumed to be at modern Hisarlık in Turkey.

ILLYRIKAN MOUNTAINS (F51, Maps 3, 4), term for part of the mountains of Europe, perhaps the Dinaric Alps in the western Balkans.

IMAOS (F69, Map 7), mountains north of INDIA, probably the central and highest part of the Himalayas.

IMBROS (F11, Map 4), island in the northeast AEGEAN, modern Gökçeada in Turkey.

INDIA (F23–4, 33–4, 37, 39, 41, 47, 49–51, 53, 59, 60, 64, 66–75, 77–8, 88, 95, 108–9, Maps 1, 2, 7), the most common toponym in the extant fragments of the *Geographika*, originally the INDOS valley, but by Eratosthenes' day the entire Indian subcontinent, essentially all of modern India and Pakistan south and east of the mountains.

INDIAN OCEAN (F70, Map 7), essentially the same as the modern term.

INDOS RIVER (F37, 64, 69–72, 74, 77–9, 81, 108, Maps 1, 7), the major river of western INDIA, which retains its ancient name.

INTERNAL SEA (F15–17, Map 3), the traditional name for the Mediterranean, although often called merely "our sea" or "the sea around us."

IONIA (F34, Map 4), district of western Anatolia.

ISSIC GULF (F13, 47, 53, Map 5), the small gulf at the northeast corner of the Mediterranean, the modern Gulf of Iskenderun.

ISTHMOS, PELOPONNESIAN (F16, Map 4), the Isthmos of Corinth.

ISTROS (F8, 15–16, 148–9, Maps 1, 3, 4), the major river of eastern EUROPE, the modern Danube.

ITALY (F6, 131, 134–5), essentially the same as the modern term, but with emphasis on the southern portions.

ITHAKA (F8, 133, Map 4), island west of Greece, the home of Odysseus, which retains its ancient name.

JUDAEA (F95, Map 5), territory of the southern Levant around Jerusalem.

KABAION (F37, Map 3), promontory in northwest France, perhaps Pointe du Raz in Brittany.

KALPE (F106, 153, Map 3), the northern PILLAR OF HERAKLES, modern Gibraltar.

KALYPSO'S ISLAND (F8, Map 3), Homeric locale, identified in antiquity with GAULOS (modern Gozo) or GAUDOS.

KANOBIC MOUTH (F37, 52, 56, 98, Map 5) of the NILE, the westernmost mouth, normally CANOPIC in Latin.

KAPPADOKIA (F88, Map 5), district of southern Anatolia, in Eratosthenes' day becoming independent from Selukid control.

KAPRIA, STRAIT OF (F6, Map 3), the channel between the Italian mainland and Capri at the south end of the Bay of Naples, modern Bocca Piccola.

KARCHEDON (F37, 60, 65, 102, 152, Maps 1, 3), the Greek term for Carthage.

KARIA (F33–4, 54, 60, 65, Maps 4, 5), territory of southwestern Anatolia, ruled by various Greek states in Eratosthenes' day.

KARMANIA (F60, 64, 77–81, 83–6, 94, Map 6), district of southern Iran opposite the mouth of the PERSIAN GULF, under loose Seleukid control in Eratosthenes' day.

KARNA or KARNANA (F95, Map 6), largest city of the Minaioi in southern ARABIA, modern Ma'in in Yemen.

KASION, MT. (F15–17, 96, Map 5), promontory on the Egyptian coast east of the Delta, modern Ras Qashrun.

KASPIAN (F113), alternative name for the KAUSAKOS MOUNTAINS, not generally accepted.

KASPIAN GATES (F37, 48, 51–2, 55–6, 60, 62–4, 77–80, 83–6, 108, Maps 1, 5), major pass and route from the Iranian plateau into Central Asia, lying southeast of the Caspian Sea, seen as the end of the civilized world and known to Greeks since Alexander's passage; an important point in Eratosthenes' geographical scheme, perhaps the modern Sar Darrah pass in the Elburz mountains, but not certainly located among several candidates.

KASPIAN SEA (F24, 33, 52, 108, 110, 112, Maps 1, 5), the modern Caspian Sea, believed in Eratosthenes' day to be an inlet of the EXTERNAL OCEAN.

KASPIOS (F52, 108, Map 5), probably the pass on the main route between the Black and Caspian Seas, perhaps modern Surami in Georgia.

KATABATHMOS (F100, Map 4), the "Descent," several places on the precipitous western Egyptian coast, most notably the "Great Katabathmos" at modern Sollum.

KATTABANIA (F95, Map 6), frankincense-producing region of southern ARABIA, modern Qataban.

KAUKASOS (F23, 47, 56, 60, 69, 71, 74, 78, 113, Maps 1, 5), specifically the region between the Black and Caspian Seas, dominated by the KAUKASOS MOUNTAINS, which can be used loosely to refer to the various mountains extending east of the Caspian Sea to north of INDIA, as far as the Himalayas.

KELTIKA (F37, 131, 153, Map 3), loose term for northwest EUROPE, the region north of IBERIA, essentially modern France.

KENCHREAI (F16, Map 4), the seaport on the Saronic Gulf side of the ISTHMOS of Corinth.

KEPHISSIAN SPRINGS (F2, Map 4), Homeric locale (*Iliad* 2.523), one of the sources of the Kephisos River in Phokis, near LILAIA.

KERKINA (F104–5, Map 3) and KERKINITIS (F105, Map 3), two adjoining islands off the African coast, modern Kerkennah Islands in Tunisia. On KERKINA was the city of the same name, modern Bordj el-Marsa.

KERNE (F13, Maps 1, 3), Carthaginian trading post in West Africa, in decline in Eratosthenes' day but still used by Greeks, location unknown.

KILIKIA (F13, 47, 50–1, Map 5), district of south central Asia Minor, largely independent in Eratosthenes' day.

KINNAMOMOPHOROI (F34–5, 53, 57–8, 98, Map 6), ethnym, whose territory was the southernmost part of the inhabited world, "the land of the cinnamomum [cassia] bearers," referring not to production of the plant but to traders of it. The region is the modern Somali coast.

KIRKAION (F6, Map 3), the home of Kirke, usually identified with Monte Circeo on the Italian coast between Rome and Naples.

KLEIDES (F130, Map 5), the easternmost point of CYPRUS, modern Cape Apostolos Andreas.

KOILE SYRIA (F60, 95–6, Map 5), "Hollow Syria," specifically the upper Orontes valley in modern Syria and Lebanon, but often used generally for interior SYRIA, west of northern MESOPOTAMIA.

KOLCHIS (F13, 16, 47, 50–2, 120, 131, Map 5), region at the southeast corner of the Black Sea (essentially modern Georgia), originally associated with the Argonauts, but by Eratosthenes' time an area of Greek settlement, important for its trading contacts to the Caspian region and beyond.

KOLYTTOS (F33, Map 4), district of ATHENS, southwest of the Agora.

KOMMAGENE (F87–8, Map 5), district of southeast Anatolia, under Seleukid control in Eratosthenes' day.

KORINTHIAN GULF (F16, Map 4), the modern Gulf of Corinth.

KORKYRA (F8–9, 133, Map 4), or Kerkyra, island off the northwest coast of Greece, which came under Roman control in Eratosthenes' day, modern Corfu.

KOTEIS (F100, Map 3), district of northwest AFRICA, probably the northern part of the ATLANTIC coast, just south of modern Cape Spartel.

KRIOUMETOPON (F129, Map 4), the "Ram's Head," the southwestern promontory of CRETE, modern Cape Krion.

KYANEAI (F52, Map 4), the "Dark Rocks," rocks at the eastern end of the BOSPOROS.

KYDNOS (F13, Map 5), river of southeastern Anatolia at Tarsos, the modern Tarsos Çay.

KYMAIAN GULF (F6, Map 3), seemingly Eratosthenes' term for the Bay of Naples, although peculiar because KYME is not on the bay.

KYME (F7, Map 3), early Greek settlement on the coast just north of the Bay of Naples, closely allied with Rome in Eratosthenes' day, modern Cuma.

KYRENE, KYRENAIKA (F16, 23, 29, 59–60, 72, 101, 104, 129, Maps 1, 5), the most significant early Greek city in Africa as well as its extensive surrounding territory, lying on a geographical parallel, but despite being Eratosthenes' hometown of no particular importance in his geographical scheme, modern Shahat in Libya.

KYRNOS (F151, Map 3), the modern island of Corsica, a Roman province in Eratosthenes' day.

KYROS (F108–9, Map 5), river of the Caucasus region, part of the trade route between the Caspian and the Black Sea, the modern Mtkvari and Kura in Georgia and Azerbaijan.

LADON (F140, Map 4), river of the northwest PELOPONNESOS, retaining its ancient name.

LAKONIA (F134, Map 4), district of the southwestern Peloponnesos.

LECHAION (F16, Map 4), city on the Gulf of Corinth, the major port of Corinth.

LEMNOS (F11, Map 4), island in the northern AEGEAN, under Athenian control in Eratosthenes' day, and today retaining its ancient name.

LEUKTRA (F139, Map 4), location in southwestern Boiotia, site of Theban victory over Sparta in 371 BC, at modern Parpoungia in Greece.

LIBYA (F8, 13, 15–16, 33, 57, 98, 100, 104, 106, Maps 2, 1, 5), one of the three continents, essentially modern Africa, but always excluding EGYPT and with little comprehension of the sub-Saharan regions.

LIGYSTIKA (F134, Map 3), ancient name for the Iberian Peninsula, probably originally the territory around MASSALIA but generically southwest EUROPE; the more familiar Latin form is "Liguria."

LIXOS (F100, 107, Maps 1, 3), Carthaginian trading settlement on the West African coast, surviving into Eratosthenes' day, at the modern Leucos River in Morocco.

LILAIA (F2, Map 4), city in Phokis mentioned in the Homeric *Catalogue of Ships* (*Iliad* 2.523), near modern Kato Agoriani in Greece.

LOPADOUSA (F105, Map 3), island in the Mediterranean between Sicily and Tunisia, modern Lampedusa.

LOTOPHAGITIS (F105, Map 3), "place of the Lotus Eaters," believed by Eratosthenes to be the island of MENINX, modern Jerba off the southeast coast of Tunisia.

LYKAONIA (F60, Map 5), district of south central Anatolia, important for its location on main trade routes, under varying control in Eratosthenes' day.

LYKIA (F60, Map 5), district of southwestern Anatolia, largely under Ptolemaic control in Eratosthenes' day.

LYKOS (F83, 119, Map 5), name of two rivers. F83 refers to one in Assyria, crossed by Alexander and on a route to the interior, the modern Great Zab in Iran and Turkey. The LYKOS RIVER of F119 empties into the Black Sea and is probably the modern Kelkit Çay.

LYNX (F100, 107, Map 3), name for city in West Africa, more properly LIXOS, and probably not used by Eratosthenes.

LYSIMACHEIA (F60, Map 4), city in Thrace founded in 309 BC and an important trading center, lying on one of Eratosthenes' parallels, modern Baklaburun on the Gallipoli peninsula of Turkey.

MAIOTIC LAKE or MAIOTIS (F24, 34, 115–16, Map 4), the large bay on the north side of the Black Sea, the modern Sea of Azov.

MAKAI, PROMONTORY OF (F94, Map 6), the peninsula on the Arabian side of the entrance to the PERSIAN GULF, modern Ras Musandam in Oman.

MAKEDONIA (F11, 136–7, Map 4), district at the northwestern corner of the AEGEAN, in Eratosthenes' day controlling much of northern Greece.

MAKEDONIAN GULF (F134, Map 4), the northernmost part of the AEGEAN.

MALEIA (F134–5, Map 4), the traditional southernmost point of the Greek peninsula (Cape Tainaron to the west is actually about 7 km. farther south), modern Cape Malea.

MARIABA (F95, Map 6), major city of the Sabaioi in southern ARABIA, perhaps modern Marib in Yemen.

MARMARIDAI (F100, Map 5), waterless desert region of the interior KYRENAIKA.

MASSALIA (F34, 60, 133, Map 3), most important Greek city of the western Mediterranean, lying on one of Eratosthenes' parallels, modern Marseilles in France.

MATIENE (F15, Map 5), little-known region between ARMENIA and MEDIA, around modern Lake Urmia in Iran.

MAUROUSIA (F39, 60, 100, 107, Map 3), northwestern coastal Africa west of the Carthaginian territory, a vast region under semi-Hellenized kings in Eratosthenes' day, modern Morocco (which preserves the name) and Algeria.

MEDIA (F13, 48, 60, 78, 83, Map 5), upland territory southwest of the Caspian Sea, the historic land of the Medes, under Seleukid control in Eratosthenes' day, located in northern modern Iran.

MELANIAN (F134, Map 4) or MELAS (F11) GULF, the northwest corner of the AEGEAN, the deep bay west of the modern Gallipoli Peninsula in Turkey known as the Gulf of Saros.

MELITE (F33, Map 4), district of Athens, west of the Agora.

MENINX (F104–5, Map 3), island on the African coast, modern Jerba off the coast of Tunisia.

MEROË (F25, 34–6, 40–1, 43, 47, 50, 57–9, 68, 98, 100, Maps 1, 6), capital of AITHIOPIA on the upper Nile between the Fifth and Sixth Cataracts, at the crossing of Eratosthenes' main meridian and a parallel, the seat of a major kingdom in his day that was an important point of contact between the Ptolemaic world and interior Africa; modern Bagrawiya in Sudan.

MEROPIA (F8), fictional land created by Theopompos of Chios.

MESOPOTAMIA (F38, 63, 83, 87, 89, Map 5), ancient territory between the TIGRIS and EUPHRATES, in modern Iraq.

METAGONION (F106, Map 3), territory of northwest Africa, near the southern PILLAR OF HERAKLES.

MINAIA (F95, Map 6), city in southern ARABIA, modern Ma'in in Yemen.

MOIRIS (F15, Map 5), lake in EGYPT west of the NILE, modern Birket Qarun in the Faiyum.

MYSIA (F60, Maps 4, 5), district of northwest Anatolia, lying on one of Eratosthenes' parallels, ruled by Pergamon.

NEAPOLIS (F6, 60, Map 3), Greek city in southern Italy, allied with ROME in Eratosthenes' day, modern Naples in Italy.

NIKAIA (F60, Map 4, 5), Greek city and Hellenistic foundation in northwest Anatolia, part of the kingdom of BITHYNIA in Eratosthenes' day, modern İznik in Turkey.

NILE (F8, 10, 33–4, 37, 41, 53, 55–6, 95, 98–100, 103, Maps 1, 5, 6), the famous river of EGYPT, whose essentially straight south-to-north course was instrumental in the determination cf Eratosthenes' main meridian.

NIPHATES (F89, Map 5), mountainous region of eastern Anatolia, south of modern Lake Van in Turkey.

OCEAN (F8–9, 33, 39, 70–2, 74, 78, 81, 94–5, 100, 103, 110, 153, Map 3), Greek OKEANOS, also called the EXTERNAL SEA, the standard term for the body of water surrounding the three continents, and which was believed to be continuous. The more vague term OCEAN was probably anachronistic in Eratosthenes' time as localizations, such as ATLANTIC, were coming into use.

OCHOS (F24, 109, Map 7), river of central ASIA, location uncertain, probably one of the rivers of BAKTRIA, or a confusion with OXOS.

OGYION (F8), unknown mountain.

OGYRIS (F94, Map 6), island south of the PERSIAN GULF, location unknown.

OLYMPIA (F140, Map 4), the major Greek sanctuary in the western PELOPONNESOS.

OPIS (F83, 89), village on the TIGRIS in central MESOPOTAMIA, probably at modern Tell Mujeih' in Iraq.

ORKYNIAN (F150, Map 3), toponymic adjective of central EUROPE, exact meaning uncertain, but perhaps the Hercynian Forest of Roman times.

OROPOS (F33, Map 4), disputed territory in central Greece on the border between Attika and Boiotia, modern Skala Oropou.

ORTOSPANA (F78, 108, Map 7), town in central ASIA, perhaps at the location of model Kabul in Afghanistan.

ORTYGIA (F6, Map 3), island at SYRAKOUSAI in Sicily, now connected to the mainland in modern Syracuse in Italy.

OSTIMIOI (F37, Map 3), promontory of northwest EUROPE, location uncertain but probably one of the headlands of Brittany.

OUXISAME (F37, Maps 1, 3), island off the northwest coast of EUROPE, perhaps modern Ushant (which was a peninsula, not an island, in antiquity, but certainly preserves the name).

OXOS (F24, 108–10, Maps 1, 7), river of BAKTRIA, the modern Amu Darya, flowing into the Aral Sea.

PALIBOTHRA or PALIMBOTHRA (F22, 69, 72, 74, Maps 1, 7), the Greek form of Pataliputra, capital of the Mauryan empire of INDIA, modern Patna on the GANGES.

PANCHAIA (F8, 14), mythical land described by Euhemeros of Messene.

PANTIKAPAION (F61, Map 5), Greek city at the northern edge of the Black Sea, part of the Bosporan kingdom, at modern Kerch in the Crimea.

PAPHLAGONIA (F60, Map 5), district of northern Anatolia, lying on one of Eratosthenes' parallels, under indigenous rule in his day.

PAPHOS (F130, Map 5), city of western Cyprus, either Palai Paphos (at modern Kouklia) or Nea Paphos (at modern Kato Paphos, 16 km.

away) founded in the late fourth century BC. Both were under Ptolemaic control in Eratosthenes' day.

PARACHOATHRAS (F108, Map 5), mountain of ARMENIA.

PARAITAKENE (F78, 83, Map 5), district of the central Iranian plateau, southeast of MEDIA.

PARAPAMISIDAI or PAROPAMISIDAI (F23, 78, Map 7), district in the Hindu Kush, northwest of INDIA and north of ARACHOSIA.

PARAPAMISOS or PAROPAMISOS MOUNTAINS (F23, 69, 108, Map 7), range in central ASIA, north of INDIA, essentially the modern Hindu Kush.

PARTHYENE (F78, Map 5), district east of the Caspian Sea, on the modern Iran-Turkmenistan border, the ancestral home of the Parthians, who broke away from Seleukid control in Eratosthenes' day.

PATALENE (F74, 77. Map 7), district at the mouth of the INDOS, in modern Pakistan.

PELETHRONION (F8, Map 4), region in the vicinity of MT. PELION in THESSALY associated with the centaur Cheiron.

PELION (F6, 8, Map 4), mountain in THESSALY, which retains its ancient name.

PELOPONNESOS (F47, 60, 129, 134, 146, Map 4), the southern portion of the Greek peninsula.

PELORIAS (F6), probably modern Capo Peloro at the eastern tip of Sicily.

PELOUSIAC MOUTH (F98, Map 5), the easternmost mouth of the NILE.

PELOUSION (F15, 17, 62, Map 5), city on the Mediterranean at the PELOUSIAC MOUTH of the NILE, the normal entry point to EGYPT from the northeast, modern Tell el-Farama.

PERSEPOLIS (F83, 86, Maps 5, 6), Persian royal city, under Seleukid control but in decline in Eratosthenes' day, at modern Taht-e Gamsid in Iran.

PERSIAN GULF (F77–9, 81, 83, 89, 93–4, 98, Maps 1, 5, 6), also called the PERSIAN SEA, the large inlet extending northwest from the Indian Ocean, which retains its ancient name.

PERSIS (F60, 64, 77–8, 81, 83, 85, Map 5), district of the southern Iranian plateau, north of the PERSIAN GULF, the ancestral home of the Persians, lying on one of Eratosthenes' parallels and under Seleukid control in his day.

PETRA (F95, Map 5), the famous trading center of northern ARABIA, which retains its ancient name, in modern Jordan.

PEUKE (F148–9, Maps 4, 5), island at the mouth of the ISTROS RIVER, not identifiable today but part of the Danube delta.

PHAROS (F10, 16, 154, Map 5), Homeric locale off the coast of EGYPT, where the famous lighthouse of ALEXANDRIA was built in the early third century BC.

PHAROS (F145, Map 3), island off the Illyrian coast, part of the Illyrian kingdom in Eratosthenes' day, modern Hvar in Croatia.

PHASIS (F8, 52, 119–20, Map 5), river of the Caucasus region, flowing into the Black Sea at KOLCHIS, the modern Rioni in Georgia.

PHENEOS (F140, Map 4), city of the northern PELOPONNESOS, modern Kalyvia in Greece.

PHRYGIA (F15, 94, Map 5), district of central Anatolia, under weak Seleukid control in Eratosthenes' day.

PILLARS OF HERAKLES (F13, 15–17, 37, 47, 49, 51, 53, 55–6, 65–6, 83, 100, 106, 133, 135, 154, Map 3), the point where the Mediterranean joins the ATLANTIC. In antiquity there was debate about what constituted the PILLARS (see F106), since there were several candidates along the 50-kilometer strait between the two bodies of water, but opinion generally favored the two mountains at the east end, KALPE and ABILYX (or ABILA), modern Gibraltar and Jebel Mousa.

PLANKTAI (F106), the "Wandering Rocks" or "Clashing Rocks," Homeric locale (*Odyssey* 12.61, 23.327) associated with the Sirens, placed by some around the PILLARS OF HERAKLES.

PONTOS (F8, 13, 15–6, 23, 33, 47, 52, 60, 115–16, 119, 131, 134, Maps 5, 6), a district of north central Anatolia, in Eratosthenes' day in the early stages of consolidating its power, but to him the term consistently refers to the sea bordering the territory, the Black Sea, a term used interchangeably with the EUXINE.

PORTHMOS (F7, 51, 65, 135, Map 3), "the Strait," but when used specifically always the one between Sicily and Italy, the modern Strait of Messina.

POSEIDONIAN GULF (F6, Map 3), the bay south of the Bay of Naples, the modern Gulf of Salerno in Italy.

PRETTANIKE (F34, Maps 1, 2, 3), original Greek name for Great Britain, but soon erroneously changed to BRETTANIKE.

PROPHTHASIA (F78, 108, Map 7), city so named by Alexander in DRANGIANE, perhaps near Farah in western Afghanistan.

PROPONTIS (F15–16, 47, 50, Map 4), the sea between the HELLESPONT and BOSPOROS, the modern Sea of Marmara.

PTOLEMAIS (F41–2, 59, 95, 97, Map 6), agricultural outpost on the Red Sea founded by Ptolemaios II, exact location unknown but roughly on the same latitude as MEROË.

PTOLEMAIS (F60, Map 5), Phoenician city named by Ptolemaios II, which became the major Hellenistic port for the southern Levant, modern Akko in Israel.

PYRENAIOI or PYRENES (F51, 133, 152, Map 3), the westernmost portion of the European mountains, still only vaguely known in Eratosthenes' day, the modern Pyrenees.

RED SEA (F41), Roman term for the western gulf of the ERYTHRAIAN SEA, the modern Red Sea.

RHINOKOLOURA (F96, Map 5), town on the Mediterranean coast of eastern Egypt, modern el-Arish.

RHIPAIA MOUNTAIN (F8, Map 3), the "tossing mountain," early term for the mountains of the far north, more mythological than an actual place.

RHODES (F34, 47, 51, 53–4, 56, 60, 65, 83, 128, 148, Maps 1, 4), island in the eastern Mediterranean and important commercial state in Eratosthenes' time, crucial to his scheme as it was where his main meridian and parallel crossed.

ROME (F60, 65, Map 3), Italian city. Eratosthenes showed no interest in its developing role toward the Greek world, merely using it as a topographical point.

SABATA (F95, Map 6), city in southern ARABIA, modern Shabwa in Yemen.

SACRED PROMONTORY (Map 1): see HIERON.

SALMYDESSOS (F15, Map 4), coastal region of Thrace on the Black Sea, near modern Midye in Turkey.

SAMOSATA (F88, Map 5), royal city of KOMMAGENE in eastern Anatolia on the EUPHRATES, modern Samsat in Turkey.

SAMOTHRAKE (F11, Map 4), island and major cult center in the northern AEGEAN, which retains its ancient name.

SARDO (F151, 154, Map 3), or SARDON, island in the western Mediterranean, which became a Roman province in Eratosthenes' day, modern Sardinia.

SARDOAN SEA (F15, 127, Map 3), the sea immediately to the west of Sardinia.

SCHERIA (F8), Homeric locale, the land of the Phaiakians, believed in antiquity to be KORKYRA (modern Corfu).

SEIRENAI or SEIRENOUSSAI (F6–7, Map 3), rocks off the coast of Italy where it was believed that the Sirens lived, probably the islets today called li Galli south of the Sorrento peninsula, which preserve the name.

SELEUKEIA (F87, 89, Map 5), city on the TIGRIS founded by Seleukos I, the major Seleukid center of the central portion of its empire, at modern Tell Umar in Iraq.

SIDON (F60, Map 5), Phoenician city, an independent state in Eratosthenes' day, modern Saida in Lebanon.

SIKELIA (F6, 8, 16, 127, Map 3), the island of Sicily, largely under Roman control in Eratosthenes' time.

SIKELIAN SEA (F15–16, 127, Map 3), the region just east of Sicily, toward mainland Greece.

SIKELIAN STRAIT (F16, 47, 53, Map 3), often just "the Strait" (PORTHMOS), the Strait of Messina.

SINOPE (F47, 60–1, Map 5), independent Greek city on the north coast of Anatolia, lying on one of Eratosthenes' parallels, modern Sinop in Turkey.

SIRBONIS (F15, Map 5), coastal lake in eastern EGYPT, the modern Sabkhat el-Bardawil.

SKEIRONIAN ROCKS (F11, Map 4), off the coast of the Megarid, associated with the giant Skeiron, who attacked Theseus; the specific location is unknown.

SKYLLA and CHARYBDIS (F6–7, Map 3), the famous monster and whirlpool of mythology, usually localized at the north end of the Strait of Messina, the latter at the Cariddi off the eastern point of Sicily, opposite the mainland village of Scilla.

SKYTHIA (F53, Maps 1, 2, 5), specifically the region north of the Black Sea, but used generically to mean the farthest parts of the world to the north and east. Eratosthenes usually used the ethnym topographically (except at F53). The SKYTHIAN DESERT (F15) seems to refer to the coastal areas north of the mouth of the ISTROS in the modern Ukraine, which, however, are difficult to characterize as "desert."

SOUNION (F134, Map 4), the southernmost point of ATTIKA in Greece.

SOUSA (F13, 64, 83, 85–6, Map 5), ancient Elamite capital and later a Persian royal residence, under Selukid control in Eratosthenes' day and lying on one of his parallels, modern Shush in Iran. The surrounding territory was SOUSIANA (F60, 94). The SOUSIS of F91 is a variant.

STETHES (F15, Maps 4, 5), "The Breasts," place on the eastern shore of the Black Sea between SALMYDESSOS and the mouth of the ISTROS, noted for its shallows.

STYMPHALOS (F140, Map 4), city and adjoining lake in the north-eastern PELOPONNESOS.

SYENE (F34, 41, 43, 57–9, 98, Maps 1, 6), Egyptian town at the First Cataract of the NILE, whose location at the tropic was essential to Eratosthenes' scheme.

SYMPLEGADES (F106, 117, Maps 4, 5), "The Clashers," a hazard to sailors, generally placed at the entrance to the Black Sea (perhaps the same as the KYANEAI), but believed by some to be at the PILLARS OF HERAKLES.

SYNORMADES (F118), "The Movers," perhaps the same as the SYMPLEGADES.

SYRAKOUSAI (F6, 60, Map 3), Greek city on the east coast of Sicily, sacked by ROME late in Eratosthenes' life, the modern Syracuse (Siracusa) in Italy.

SYRIA (F13, 60, Map 5), district of the Levant, stretching south from Anatolia to Phoenicia and toward the interior, in Eratosthenes' day the heart of the Seleukid empire, loosely corresponding to modern Syria.

SYRRENTON (F6, Map 3), city on the south edge of the Bay of Naples, an Oscan outpost, modern Sorrento in Italy.

SYRTES (F100, 104–5, Maps 3, 4), the great bays of the Mediterranean in northeast AFRICA. The eastern portion, which includes the southernmost point of the Mediterranean and lies just west of the KYRENAIKA, is the Greater Syrtis (the modern Gulf of Sidra off Libya); to the west is the Lesser Syrtis (the modern Gulf of Gabès off Tunisia).

TAGOS (F133, Map 3), river of IBERIA, the modern Tajo (Spain) or Tejo (Portugal.

TAMASSOS (F130, Map 5), mining location on CYPRUS, in the central uplands near modern Politiko.

TAMNA (F95, Map 6), city of southwest ARABIA, modern Hajar Kuhlan in Yemen.

TANAIS (F8, 14, 24, 33, 38, 135, Map 5), river emptying into the northeastern Black Sea, the most remote toponym in the northeast of the inhabited world, the modern Don.

TAPROBANE (F35, 53, 57, 74, 76, Maps 1, 7), island to the south of INDIA, little known in Eratosthenes' day, modern Sri Lanka.

TARRAKON (F152, Map 3), indigenous city on the northeast coast of IBERIA, under Roman control late in Eratosthenes' day, modern Tarragona in Spain.

TARTESSIS (F153, Map 3), district of southern IBERIA, important trading area under Phoenician and later Carthaginian control, with

GADEIRA as its most important city, essentially the southwestern part of modern Anadalucía in Spain.

TAUROS (F47–9, 53, 55–6, 63, 66, 69, 71–2, 77, 87–8, Maps 1, 5), east-west mountain range of ASIA, seen by Eratosthenes as dividing the inhabited world, originally the mountains of southern and eastern Anatolia (where the name survives), but also used to describe the mountains east through Iran (the modern Elburz) and into Afghanistan as far as the Hindu Kush.

TEREDON (F83, 94, Map 5), village in lower Mesopotamia near the coast, location unknown because of changes in topography, but perhaps near modern Basra in Iraq.

TERSOS (F124, Map 5), early name for Tarsos, the ancient trading city of southeastern Anatolia.

THAPSAKOS (F52, 55–6, 62–3, 80, 83–4, 87, 94, Map 5), ancient major crossing point on the EUPHRATES, used by Alexander, and an important point in Eratosthenes' scheme, location unknown, perhaps near modern Dibse in Syria.

THASOS (F11, Map 4), island in the northern AEGEAN, a Makedonian possession in Eratosthenes' day, retaining its ancient name.

THERMAIC GULF (F65, Map 4), at the northwest corner of the AEGEAN, which retains its ancient name.

THERMODON (F8, 119, Map 5), river of northern Anatolia, emptying into the Black Sea, perhaps the modern Terme Çay in Turkey.

THESSALONIKE (F133, Map 4), Makedonian port city, retaining its modern name.

THESSALY (F18, Map 4), district of northern Greece. It is not certain that Eratosthenes mentioned the region.

THISBE (F2, 8, Map 4), city in western Boiotia mentioned in the Homeric *Catalogue of Ships* (*Iliad* 2.502), at modern Kakkosi.

THOAS (F105, Map 3), town on the island of MENINX off the North African coast, location uncertain but probably at modern Houmt Souq in Tunisia.

THOPITIS (F87, 89, Map 5), lake that the TIGRIS allegedly flows through, probably modern Lake Van in Turkey, which, however, is not in the TIGRIS drainage.

THOULE (F14, 34–5, 37, Maps 1, 2, 3), place in the far north discovered by Pytheas of MASSALIA but not reached by anyone else, the most northern place in the world, almost certainly Iceland.

THRACE (F11, 136, Map 4), district to the north of the AEGEAN, under Makedonian control in Eratosthenes' day.

THRACIAN CHERSONESOS (F134, Map 4), the peninsula at the east end of THRACE between the HELLESPONT and the MELANIAN GULF, the modern Gallipoli Peninsula in Turkey.

THRACIAN MOUNTAINS (F51, Map 4), term used generally for the mountains north of the AEGEAN, such as the modern Rhodope range on the Greek-Bulgarian border.

THYREA (F33, Map 4), plain on the west side of the Gulf of Argos, disputed by Argos and Sparta, the modern Kynouria district in Greece.

TIGRIS (F13, 38, 63, 83, 87, 89, Map 5), the eastern river of MESOPOTAMIA.

TINX (F100, Map 3), indigenous name for LIXOS, on the West African coast.

TITENIS (F121), district and river around the Black Sea, location unknown.

TOMISA (F88), town on the upper EUPHRATES, probably not mentioned by Eratosthenes, location uncertain but perhaps near modern İzolü in Turkey.

TROAD (F34, Map 4), district of northwest Anatolia.

TROGODYTIKA (F41, 59, 95, 97–8, Map 6), district between the Red Sea and NILE, mostly in modern Sudan, loosely allied with Ptolemaic EGYPT. Ancient and modern writers often misspell the toponym as "Troglodytika," which gives an erroneous impression.

TROY (F2, Map 4), famous ancient city in northwest Anatolia, probably at modern Hisarlık in Turkey.

TYRE (F60, Map 5), Phoenician city, under Ptolemaic control in Eratosthenes' day, modern es-Sur in Lebanon.

TYROS (F94, Map 6) or TYLOS, island in the PERSIAN GULF, of interest to Greeks because of its exotic tropical flora and fauna, modern Bahrain.

TYRRHENIA (F6, Map 3), the Greek term for the land of the Etruscans.

TYRRHENIAN SEA or GULF (F16, 127, 134, Map 3), the sea between Italy, Corsica, Sardinia, and Sicily.

WALL OF SEMIRAMIS (F83, 89, Map 5), traditional name for a fortification running between the TIGRIS and EUPHRATES where they were closest together (in antiquity about 40 km. apart), more commonly called the "Wall of Media," and believed to be the existing wall remains in the Habl es-Sahr region of Iraq.

WESTERN SEA (F16), early term for the ATLANTIC OCEAN.

XANTHOS (F60, Maps 4, 5), primary city of LYKIA in Anatolia, under Ptolemaic control in Eratosthenes' day.

ZARIASPA (F108, Map 7), primary city of BAKTRIA, also known as BAKTRA, modern Balkh in Afghanistan.

ZEUGMA (F87–8, Map 5), "The Crossing," major crossing of the EU-PHRATES in KOMMAGENE, carefully controlled by the Selukids, modern Balkis in Syria.

ZONOS RIVER (F111, Map 5), affluent into the Caspian Sea, perhaps the modern Ural.

MAPS

250

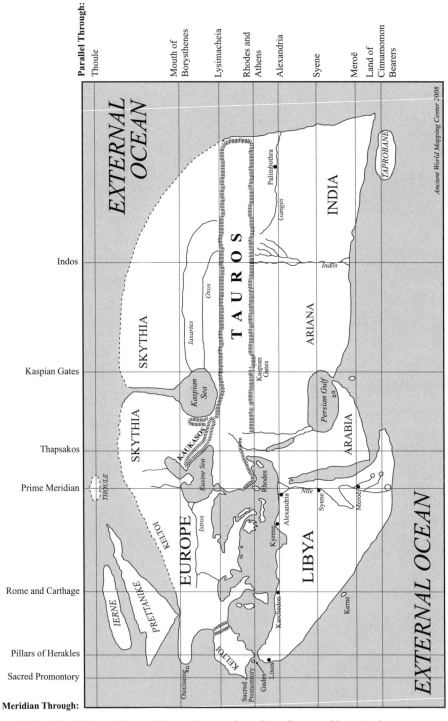

Thoule
Mouth of Borysthenes
Lysimacheia
Rhodes and Athens
Alexandria
Syene
Meroë
Land of Cinnamomon Bearers

EXTERNAL OCEAN

Ancient World Mapping Center 2008

TAPROBANE

Palimbothra

INDIA

Ganges

Indos

TAUROS

SKYTHIA

Iaxartes

Oxos

ARIANA

Indōs

Kaspian Gates

Kaspian Gates

Kaspian Sea

Persian Gulf

ARABIA

SKYTHIA

Thapsakos

KAUKASOS

Euxine Sea

Rhodes

Nile

Prime Meridian

THOULE

Alexandria

Syene

Meroë

EUROPE

Istros

KELTOI

Kyrene

LIBYA

IERNE

PRETTANIKE

Karchedon

Rome and Carthage

Kerne

EXTERNAL OCEAN

KELTOI

Pillars of Herakles

Ouxisame

Sacred Promontory

Sacred Promontory

Gades

Lixos

Indos

Kaspian Gates

Thapsakos

Prime Meridian

Rome and Carthage

Pillars of Herakles

Sacred Promontory

Map 1. The World According to Eratosthenes (based on *Grosser Historischer Weltatlas* [fifth edition, Munich 1972], p. 12, and redrawn by the Ancient World Mapping Center).

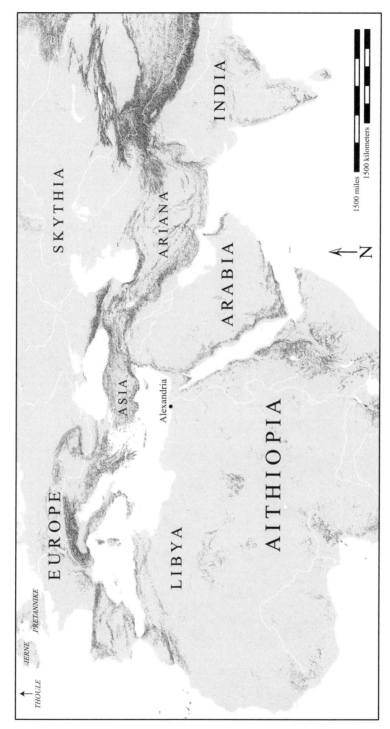

Map 2. The Known World in the Second Century BC (based on *Grosser Historicher Weltatlas* [fifth edition, Munich 1972], p. 13, and redrawn by the Ancient World Mapping Center).

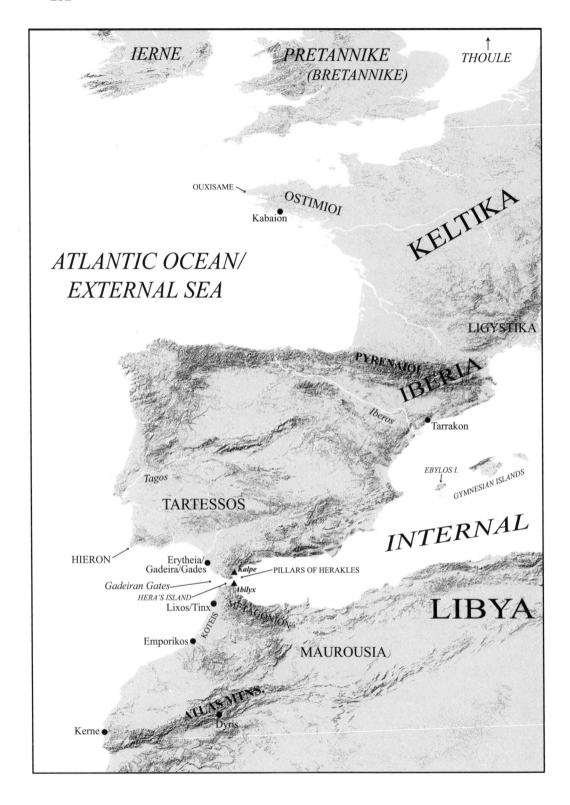

IERNE

PRETANNIKE
(BRETANNIKE)

THOULE

ATLANTIC OCEAN/
EXTERNAL SEA

OUXISAME

OSTIMIOI

Kabaion

KELTIKA

LIGYSTIKA

PYRENAIOI

IBERIA

Iberos

Tarrakon

EBYLOS I.

GYMNESIAN ISLANDS

Tagos

TARTESSOS

INTERNAL

HIERON

Erytheia/
Gadeira/Gades

Kalpe

PILLARS OF HERAKLES

Gadeiran Gates

Abilyx

HERA'S ISLAND

Lixos/Tinx

METAGONION

LIBYA

KOTEIS

Emporikos

MAUROUSIA

ATLAS MTNS.

Kerne

Dyris

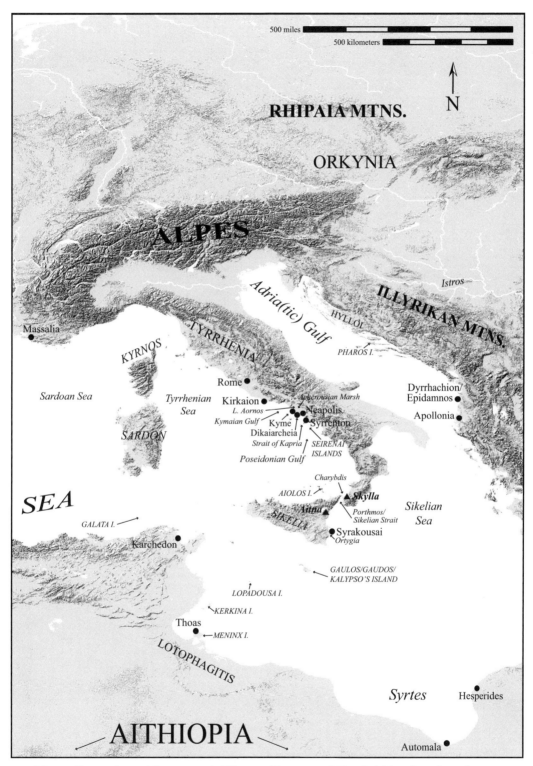

Map 3. Eratosthenes' Toponyms in Western Europe and North Africa (based on the appropriate maps in *BA* and drawn by the Ancient World Mapping Center).

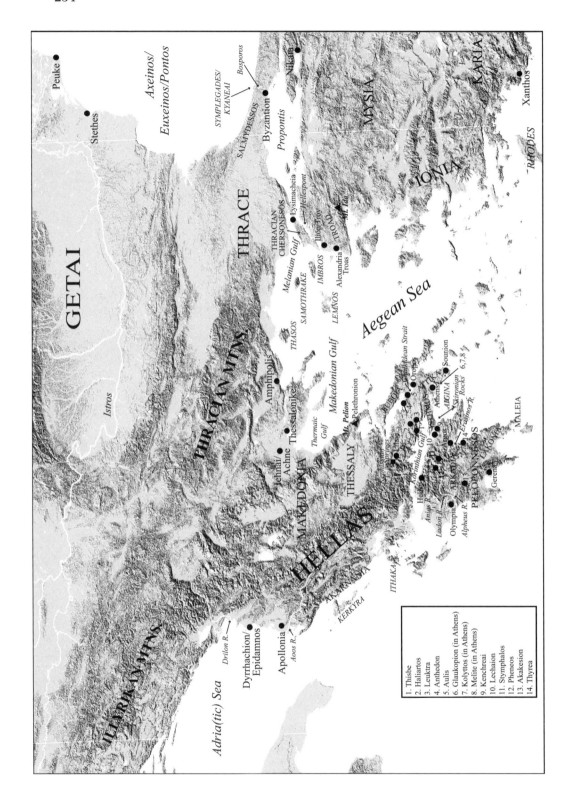

GETAI

Peuke

Stethes

Axeinos/

Euxeinos/Pontos

Bosporos

*SYMPLEGADES/
KYANEAI*

Nikaia

SALMYDESSOS

Byzantion

Propontis

THRACE

THRACIAN
CHERSONESOS

Hellespont

Lysimacheia

Ilion/Troy

Mt. Ida

TROAD

Alexandria
Troas

MYSIA

KARIA

Xanthos

IONIA

RHODES

Melanian Gulf

THASOS

SAMOTHRAKE

IMBROS

LEMNOS

Aegean Sea

THRACIAN MTNS

Istros

Amphipolis

Thessalonike

MAKEDONIA

Jonna/
Achne

*Thermaic
Gulf*

Makedonian Gulf

Mt. Pelion

Pelethronion

THESSALY

HELLAS

Aegean Strait

EUBOIA

Oropos

Athens

AIGINA

*Skironian
Rocks*

Sounion

6,7,8

Erasinos R.

Korinthian Gulf

Hellas

Arias R.

Ladon R.

Olympia

Alpheus R.

PELOPONNESOS

LAKONIA

MALEIA

Geren

Apropos

AKARNANIA

ITHAKA

KERKYRA

ILLYRIAN MTNS

Adria(tic) Sea

Drilon R.

Dyrrhachion/
Epidamnos

Apollonia

Aoos R.

1. Thisbe
2. Haliartos
3. Leuktra
4. Anthedon
5. Aulis
6. Glaukopion (in Athens)
7. Kolyttos (in Athens)
8. Melite (in Athens)
9. Kenchreai
10. Lechaion
11. Stymphalos
12. Pheneos
13. Akakesion
14. Thyrea

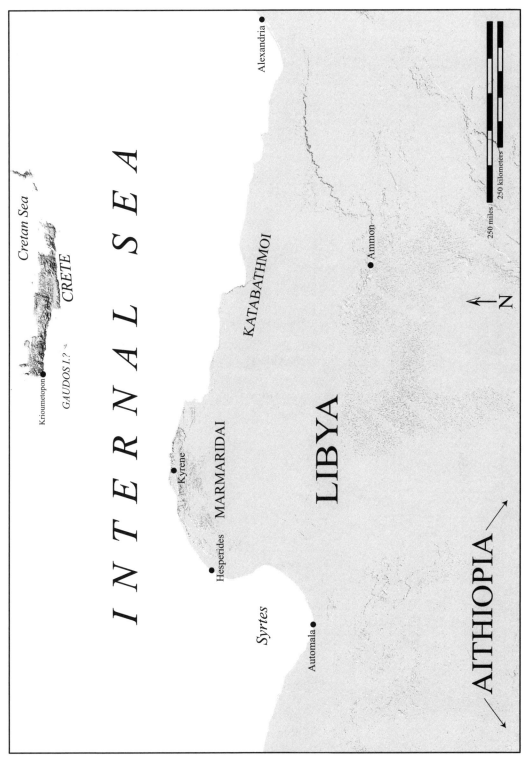

Map 4. Eratosthenes' Toponyms in the Greek Peninsula, Western Asia Minor, and Central North Africa (based on the appropriate maps in *BA* and drawn by the Ancient World Mapping Center).

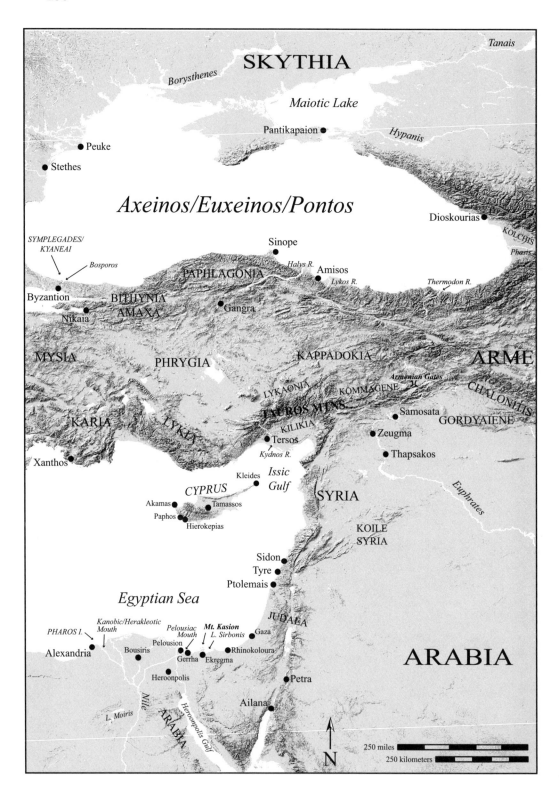

Tanais

SKYTHIA

Borysthenes

Maiotic Lake

Pantikapaion ●

Hypanis

● Peuke

● Stethes

Axeinos/Euxeinos/Pontos

Dioskourias ●

KOLCHIS

Phasis

*SYMPLEGADES/
KYANEAI*

Sinope ●

Bosporos

Halys R. Amisos ●

PAPHLAGONIA

Lykos R.

Thermodon R.

Byzantion ●

BITHYNIA
AMAXA

Gangra ●

Nikaia ●

MYSIA

PHRYGIA

KAPPADOKIA

ARME

LYKAONIA

Armenian Gates

KOMMAGENE

CHALONITIS

TAUROS MTNS

Samosata ●

GORDYAIENE

KARIA

LYKIA

KILIKIA

Tersos ●

Zeugma ●

Kydnos R.

Thapsakos ●

Xanthos ●

Kleides ●

*Issic
Gulf*

Euphrates

CYPRUS

Akamas ● ● Tamassos

SYRIA

Paphos ●
Hierokepias ●

KOILE
SYRIA

Sidon ●

Tyre ●

Ptolemais ●

Egyptian Sea

JUDAEA

*Kanobic/Herakleotic
Mouth*

*Pelousiac
Mouth*

Mt. Kasion
L. Sirbonis

Gaza ●

PHAROS I.

Pelousion ●

Bousiris ●

● Rhinokoloura

Alexandria ●

Gerrha ● Ekregma

ARABIA

Heroonpolis ●

Petra ●

Nile

ARABIA

L. Moiris

Heroonpolis Gulf

Ailana ●

N

250 miles

250 kilometers

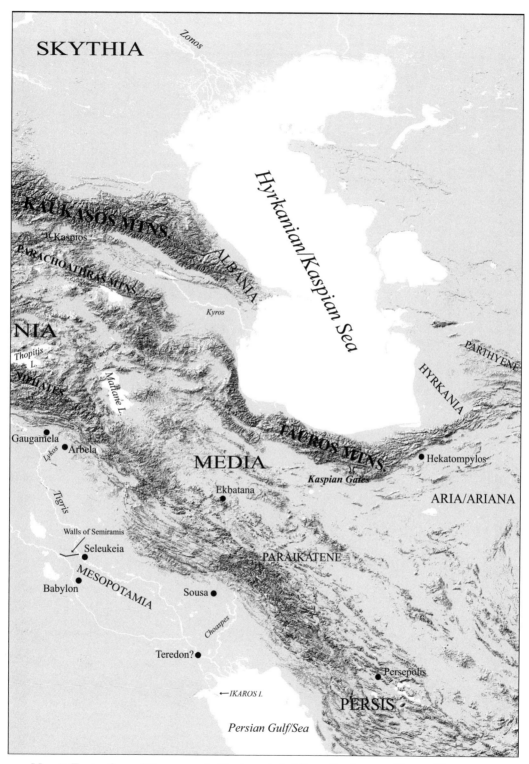

SKYTHIA

Zonos

KAUKASOS MTNS

Kaspios

PARACHOATHRAS MTNS

ALBANIA

Hyrkanian/Kaspian Sea

Kyros

NIA

Thopitis L.

NIPHATES

Marhanē L.

PARTHYENE

HYRKANIA

TAUROS MTNS

Gaugamela

Lykos ● Arbela

MEDIA

● Hekatompylos

Kaspian Gates

● Ekbatana

ARIA/ARIANA

Tigris

Walls of Semiramis

PARAIKATENE

Seleukeia

MESOPOTAMIA

Babylon   Sousa ●

Choaspes

Teredon? ●

←*IKAROS I.*

● Persepolis

PERSIS

Persian Gulf/Sea

Map 5. Eratosthenes' Toponyms in Eastern Asia Minor, the Levant, Egypt, Mesopotamia, and the Iranian Plateau (based on the appropriate maps in *BA* and drawn by the Ancient World Mapping Center).

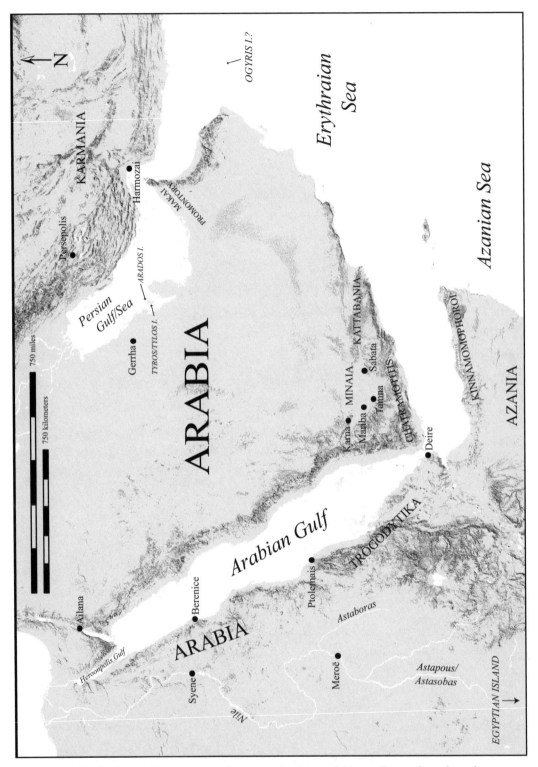

Map 6. Eratosthenes' Toponyms in Southern Arabia and Upper Egypt (based on the appropriate maps in *BA* and drawn by the Ancient World Mapping Center).

259

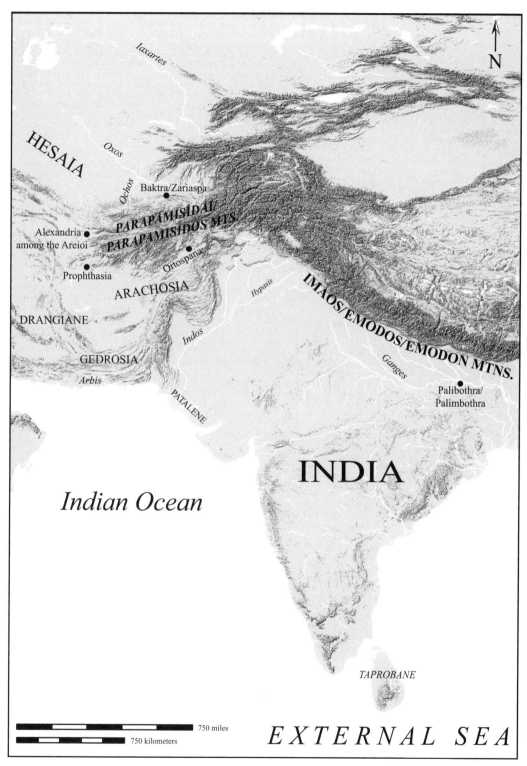

Map 7. Eratosthenes' Toponyms in the Eastern Part of the Known World (based on the appropriate maps in *BA* and drawn by the Ancient World Mapping Center).

.

APPENDICES

Appendix 1
On the Measurement of the Earth

Early commentators such as Bernhardy and Berger did not recognize the existence of this work, but included its fragments in Book 2 of the *Geographika*. For a full discussion of why the *Measurement* is indeed a separate work, see A. Thalamas, *Le géographie d'Ératosthène* (Versailles 1921) 65–78. For further on the treatise, see also G. Knaack, "Eratosthenes von Kyrene" (#4), *RE* 6 (1907) 364–6; and for discussions on the calculations involved, see *Cleomedes* (ed. Bowen and Todd), figs. 14–15 and pp. 81–4; Kidd, *Commentary* 720–2; Geus, *Eratosthenes* 223–59; Georgio Dragoni, *Eratostene e l'apogeo della scienza greca* (Bologna 1979) 161–232.

There are three testimonia that identify the *Measurement*:

a. Heron of Alexandria, *Dioptra* 35, where the title is quoted: Περὶ τῆς ἀναμετρήσεως τῆς γῆς.
b. Galen, *Institutes of Logic* 26–7, a summary of mathematical and astronomical issues, all from a single work of Eratosthenes', with no title provided, but not from the *Geographika*.
c. Macrobius, in his *Commentary on the Dream of Scipio* (1.20.9), provided a title, "Eratosthenes in libris dimensionum," which seems a Latin condensation of Heron's title. The topic is the relative size of the sun and earth.

The fragments of the *Measurement* follow:

M1 (IIB19). Geminos, Introduction to Astronomy *16.6–9.*
(6) In order to measure carefully the great circle of the earth below the celestial meridian, finding that it is 252,000 stadia with a diameter of 84,000 stadia, one divides the celestial meridian into 60 parts, with each section called a sixtieth and containing 4,200 stadia, for if 252,000 stadia are divided by 60, a sixtieth is 4,200 stadia.

(7) The distances between the zones are determined in the following manner. The two cold zones each have a width of six sixtieths, which is 25,200 stadia. The two temperate zones each have a width of five sixtieths, which is 21,000 stadia. The hot zone has a width of eight sixtieths, so that from the equator to the tropic on either side is four sixtieths, which is 16,800 stadia.

(8) From the terrestrial pole, which lies below the celestial pole, to the terrestrial arctic [circle], is therefore 25,200 stadia, and from the terrestrial arctic, which lies under the celestial arctic, to the terrestrial tropic, which lies under the celestial summer tropic, is 21,000 stadia. From the summer tropic to the terrestrial equator, which lies under the celestial equator, is 16,800 stadia. (9) In turn, from the equator to the other tropic is 16,800 stadia, from the tropic to the arctic is 25,200 stadia, and from the arctic to the other pole, 25,200. All together, the distance between the poles is 126,000 stadia, which is half the circumference of the earth. Thus from pole to pole is a semicircle.

M2 (IIB24). Macrobius, Commentary on the Dream of Scipio *2.6.2–5.*
(2) The whole circuit of the earth, that is, the circuit that encloses the entire circumference, represented by ABCD, has been divided by those who calculated its measurements into 60 parts. (3) The entire circumference is 252,000 stadia, and thus each part extends for 4,200 stadia. Thus, without doubt, its midpoint, that is from D east through A as far as C, has 30 [parts], and 126,000 stadia. A quarter portion, that is from A to C, beginning at the middle of the torrid [zone], has 15 [parts] and 63,000 stadia. By providing the measurement of this quarter portion, the entire dimension of the whole circumference can be established. (4) From A to N, which is in the middle of the torrid [zone], is four sixtieths, which makes 16,800 stadia. Therefore the entire torrid zone is eight sixtieths and contains 33,600 stadia. (5) The width of our zone, which is the temperate one, that is from N to I, is five sixtieths, which makes 21,000 stadia. The extent of the frigid [zone], that is from I to C, is six sixtieths, which is 25,200 stadia.

M3 (IIB30). Vitruvius 1.6.9.
... the circumference of the world, determined by means of the path of the sun and the equinoctial shadows of the gnomon and the inclination of the sky, was discovered by Eratosthenes of Kyrene through mathematical calculations and geodesic methods to be 252,000 stadia. ...

M4 (IIB31). Pliny, Natural History *2.247.*
Eratosthenes (perceptive in everything cultural, but especially knowledgeable in this matter, and who, I see, is accepted by all) established its entire circumference as 252,200 stadia, which in Roman measurements is calculated as 31,500 miles. This is a daring presumption, but expressed by such a subtle argument that one is ashamed not to believe.

M5. Censorinus, On the Birthday *13.2.*
For Eratosthenes, through geodesic calculation, concluded that the maximum circumference of the earth was 252,000 stadia.

M6 (IIB34, 35). Kleomedes, Elementary Theory *1.7.*
Natural philosophers have had many theories about the size of the earth, but the better ones are those of Poseidonios and Eratosthenes. The latter proves its size by a geodesic method, but that of Poseidonios is simpler. Each takes certain assumptions and then comes to a proof by means of the result of the arguments.

[Poseidonios' method deleted.]

That of Eratosthenes uses a geodesic method and seems somewhat obscure [some general assumptions deleted]. He says that Syene and Alexandria lie below the same meridian. Since the meridians are great circles in the cosmos, those of the earth lying below them necessarily are also great circles. Thus this procedure shows that the size of the circle of the earth through Syene and Alexandria will be the same as the great circle of the earth. He says, and this is so, that Syene lies below the summer tropical circle. Thus when the sun enters the Crab and makes the summer solstice, exactly at this meridian, by necessity the gnomons on the sundials are shadowless, since the sun lies exactly above. This is said to be [an area] 300 stadia in diameter. In Alexandria, however, at the same time, the gnomons on the sundials cast a shadow, since this city lies further north than Syene. Since the cities lie below a meridian and a great circle, if we draw an arc from the point of the shadow of the gnomon to the base of the gnomon on the sundial at Alexandria, this arc will be a section of the great circle in the basin of the sundial, since the basin of the sundial lies below a great circle. If we then conceive of straight lines falling through the earth from each of the gnomons, they will come together at the center of the earth. Since the sundial at Syene lies directly below the sun, if we further conceive of a straight line going from the sun to the point of the gnomon of the sundial, it will be a single straight line going from the sun to the center of the earth. If we conceive of another from the point of the shadow of the sundial going from the basin of the sundial at Alexandria up to the sun, this and the previous line will be parallel, extending from different parts of the sun to different parts of the earth. Another straight line extending from the center of the earth to the gnomon at Alexandria meets these parallel lines, and thus makes the alternate angles equal. One of these is at the center of the earth at the intersection of the lines drawn from the sundials to the center of the earth. The other is where the point of the gnomon at Alexandria intersects that drawn from the point of the shadow of the gnomon to the sun through the point where the line touches.

The arc drawn from the point of the shadow of the gnomon around to its base stands on this, and that from Syene to Alexandria stands on the center of the earth. The arcs are similar to each other, since they stand on equal angles. The ratio of the one in the base to its own circle is the same as the ratio of that drawn from Syene to Alexandria. The one in the basin is found to be a fiftieth part of its own circle. Necessarily, then, the distance from Syene to Alexandria must be a fiftieth part of the great circle of the earth, and this is 5,000 stadia. Thus the entire circle is 250,000 stadia. This is Eratosthenes' procedure.

At the winter solstices sundials are placed in each of these cities, and when each casts a shadow, necessarily the one at Alexandria is found to be longer, because that city is farther from the winter tropic. Taking the amount that the shadow at Syene is exceeded by that at Alexandria, they find this to be a fiftieth part of the great circle of the sundial. Thus from this it is also obvious that the great circle of the earth is 250,000 stadia.

M7 (IIB41). *Martianus Capella 6.596–8.*

The circumference of the earth was calculated at 252,000 stadia by the most learned Eratosthenes, using a gnomon. There are round containers of bronze called *skaphia*, which mark the course of the hours by means of a vertical stylus placed in the center. This stylus is called a gnomon, whose shadow, measured at the equinox by determining its distance from the center when multiplied 24 times provides the measure of a double circle. Eratosthenes was provided by the surveyors of King Ptolemaios the number of stadia from Syene to Meroë, and discovered what portion of the earth this was and multiplied it by the appropriate amount, determining without hesitation how many thousands of stadia make up the circumference of the earth.

M8 (IIB42). Ptolemy, Mathematical Syntaxis *1.67.22*
I have found that the arc from the northernmost limit to the southernmost, that is the section between the tropics, is always 47 degrees and more than two thirds of a degree but less than three quarters, which is almost the same as what Eratosthenes calculated and Hipparchos also used, for the arc between the tropics is almost exactly 11 out of the 83 in the meridian.

M9 (IIB42). Theon of Alexandria, Commentary on the Mathematical Syntaxis *(ed. Rome, vol. 2, p. 528, 20).*
This is almost the same as that of Eratosthenes, which Hipparchos also used, because it had been calculated accurately, as Eratosthenes calculated the entire circle at 83, and found that the part between the tropics was 11. And 360 is to 47 degrees 42 minutes 40 seconds as 83 is to 11.

Appendix 2
Testimonia for the Life of Eratosthenes

The sources for the life of Eratosthenes are scattered. See the biographical sketch in this volume (supra, pp. 7–15) and *FGrHist* #241, T1–14. Following are the five most important.

1. Souda, *"Eratosthenes."*

Eratosthenes, son of Aglaos, or of Ambrosios, a Kyrenaian, a student of the philosopher Ariston of Chios, the philologist Lysanias of Kyrene, and the poet Kallimachos. He was summoned from Athens by the third Ptolemaios and survived until the fifth. Because he was second in everything that he knew, although approaching the best in his education, he was called "Beta," and by others the second or new Plato. Others called him Pentathlos. He was born in the 126th [?] Olympiad [276–273 BC] and died at the age of 80, refusing food because of his weakening eyesight. He left as his most distinguished pupil Aristophanes of Byzantion, whose pupil was in turn Aristarchos [of Samothrake]. His [Eratosthenes'] pupils included Mnaseus, Menandros, and Aristis. He wrote philosophy, poetry, and history, *Astronomy* or *The Constellations*, *On the Systems of Philosophy*, *On Living Without Pain*, many dialogues, and extensive philological works.

Critical Note: The date of Ol. 126 is generally rejected, as it would be too late for study with Zenon of Kition (Strabo 1.2.2, #3 below). Either of the two previous Olympiads is possible (and each would require the change of only a single letter), with Ol. 124 (284–281 BC) perhaps the better choice (see Fraser, *PBA* 176).

2. Oxyrhynchos Papyrus *#1241, col. 2, lines 1–8.*

. . . [Apollonios] the son of Silleus, of Alexandria, called the Rhodian, a student of Kallimachos and the teacher of the first [third?] king. Eratosthenes succeeded him, and after him Aristophanes the son of Apelles, of Byzantion, and then Apollonios of Alexandria, called the "Classifier," and after him Aristarchos son of Aristarchos, originally from Samothrake: he became the teacher of the children of [Ptolemaios VIII] Philopator [perhaps an error for Ptolemaios V or VI]. After him was Kydas, from the spearman, and under the ninth king [Ptolemaios VIII?] there flourished Ammonios, Zenodotos, Diokles, and Apollodoros the grammarian.

Critical Note: The papyrus, from the second century AC, is a catalogue of lists of historical and mythological data. It is the primary source for the sequence of Librarians. See Bernard P. Grenfell and Arthur S. Hunt, *The Oxyrhynchus Papyri* 10 (London 1914) 99–112.

3. Strabo 1.2.2.

As he [Eratosthenes] himself says, he knew notable men. "For," he says, "it happened at this particular time, as never before within one enclosure or one city, that philosophers flourished, such as those around Ariston and Arkesilaos." I do not think that this is enough, as there must be a precise definition as to which of them one should follow. But he places Arkesilaos and Ariston at the head of those who flourished in his time. Apelles is also conspicuous to him, as well as Bion, whom, he says, was the first to clothe philosophy in flowery dress, but people often said this about him:

> Bion [shows] such, from under his rags.
> *[Homer,* Odyssey *18.74].*

In these very statements he demonstrates a significant weakness in his own judgment, for, although he was acquainted in Athens with Zenon of Kition, he does not mention any of his successors, but he only mentions those who differed with him and who could not keep their own school alive and flourishing at the time. His publication titled *On the Good* shows this, as well as his *Exercises* and whatever else he wrote in such a manner: he was thus in the middle between wishing to be a philosopher and lacking the confidence to attempt this undertaking, only advancing far enough to appear to be one, having provided himself an escape for his regular cycle, whether for education or amusement. This is also the fashion in his other writings.

Critical Note: The Homeric line refers to the disguised Odysseus, substituting "Bion" and omitting the remainder of the line, which in its entirety reads "Such a thigh the old man shows from under his rags."

4. *Archimedes,* Method of Mechanical Theorems, Preface.

Archimedes to Eratosthenes, greetings. Previously I sent you some of the theorems I had discovered. . . . Moreover, I see in you, as I say, an eager student and a notably eminent scholar. . . .

   Critical Note: This is the only extant contemporary comment on Eratosthenes and demonstrates the professional relationship he had with Archimedes.

5. *Dionysios of Kyzikos* (Greek Anthology 7.78).

A softening old age with no darkening through disease quenched you and put you to deserved sleep pondering great things, Eratosthenes. Mother Kyrene did not receive you into the paternal tombs, son of Aglaos, but you are buried as a friend in a foreign land, here on the edge of the shore of Proteus.

   Critical note: Dionysios is otherwise unknown. This is probably the actual epitaph for Eratosthenes, erected at his tomb in Alexandria when he died around 200 BC. The "shore of Proteus" is a Homeric reference to Egypt (*Odyssey* 4.365, 385).

Appendix 3
Lengths of Measurement

In 1949 the distinguished scholar of ancient mathematical geography, Aubrey Diller, wrote:

> . . . the Greek stade was variable and in particular instances almost always an uncertain quantity. The most problematical aspect of the ancient measurements of the earth is the length of the respective stades. Some light can be thrown on it, but the matter requires circumspection, and those who blithely convert in casual parentheses or footnotes are usually unaware of the difficulties and mistakes in their statements.[1]

Nothing has changed in the past 60 years. The length of the stadion is as uncertain as ever, yet many scholars continue to ignore Diller's wise advice.[2] Although those who attempt to calculate the length of Eratosthenes' stadion are generally concerned with the *Measurement* rather than the *Geographika*, they rarely take into account another essential fact about Eratosthenes' scholarship: there is no reason to believe that Eratosthenes always used the same length of stadion. In fact, he could not. Most of his data was based on overland or oversea distances obtained

---

[1] Aubrey Diller, "The Ancient Measurements of the Earth," *Isis* 40 (1949) 7–8.

[2] See, for example, Massimo Cimino, "A New, Rational Endeavour for Understanding the Eratosthenes Numerical Result of the Earth Meridian Mesurement," in *Compendium in Astronomy* (ed. Elias G. Mariolopoulos et al., Dordrecht 1982) 11–21, who determined Eratosthenes' stadion to be 157.5 m.; Donald Engels, "The Length of Eratosthenes' Stade," *AJP* 106 (1985) 298–311, at 184.98 m., but with serious reservations; and Edward Gulbelkian, "The Origin and Value of the Stadion Unit Used by Eratosthenes in the Third Century B.C.," *AHES* 37 (1987) 359–63, at 166.7 m. See also Sarah Pothecary's astute "Strabo, Polybios, and the Stade," *Phoenix* 49 (1995) 49–67, which wisely avoids considering Eratosthenes. If one takes the difference of 27.5 m. between the above measurements and applies it to Eratosthenes' 23,000 stadia from Borysthenes to Meroë (F36), it becomes 632.5 km., hardly inconsequential.

from travelers' and sailors' reports, not astronomy. Moreover, Eratosthenes used several additional forms of measurement: the *schoinos* (F27, 72), the sailing day (F94), and the caravan day (F95). And to complicate matters further, many of his distances survive only in Roman miles (F27–8, 42, 70, 93, 97, 101, 103, 105, 111, 115–16), which he never used.

The *stadion* first appears in Greek literature in the fifth century BC. Herodotos was probably the first to cite it: he defined it as 600 feet.[3] Its use by Aristophanes indicates that it was firmly in the vernacular by the beginning of the fourth century BC.[4] But the fact remains that just as there was no consistent scheme of chronology, there was none for distances and measurements, and ancient writers were often as confused as modern ones. A metrological table of late antiquity, attributed to Julian of Askalon, calculates the Roman mile at 8¼ of the stadia Eratosthenes and Strabo used, adding that the equivalent of "today" is 7½ stadia.[5] Yet Strabo himself wrote that "most" calculate eight stadia to the mile, but Polybios used a stadion that equaled 8⅓ to the mile.[6] Pliny used a conversion of eight stadia to the mile for Eratosthenes (F27). Using the equivalent of 1,480 m. to the Roman mile, this provides a range of 177.7 to 197.3 m. for a stadion. This may seem slight until one takes into account the vast distances Eratosthenes was using, which demonstrates part of the problem: before the Hellenistic period few had any interest in tens of thousands of stadia, so minor differences in length were unimportant. Eratosthenes' other measurements are even more uncertain: sailing and caravan days are anyone's guess, and the *schoinos* was said to be 40 stadia for Eratosthenes (F27) but could vary between 30 and 120 stadia even over the same route,[7] a revealing indicator of the pitfalls of understanding ancient measurements.

Yet the important point is that, given these variables, and doubtless others that are unknown, it strains credulity to believe that one can determine the actual length of each and every of the many stadion distances recorded by Eratosthenes. It would have been impossible for him to have used stadia of the same length throughout. His distances were acquired from a variety of sources over a century, from Pytheas and the Alexander companions (if not earlier) to his own time. More importantly, they covered a wide geographical range: from eastern India to East Africa to Central Asia and northwest Europe. There is no way of determin-

[3] Herodotos 1.26, 2.149, etc.
[4] Aristophanes, *Frogs* 91; *Clouds* 430.
[5] See Poseidonios, F203 Kidd.
[6] Strabo 7.7.4.
[7] Strabo 17.1.24.

ing the degree of accuracy of Eratosthenes' informants, or whether stadion distances published by these sources had already been converted
from other measurements, and how accurately. One suspects that many
of Eratosthenes' sources provided data in *schoinoi* (something implied
in F27) and that he converted these, obviously at 40 stadia to a *schoinos*:
but there is no guarantee that the original *schoinoi* were all of the same
length. It is unlikely that Eratosthenes' sources gave equivalents or defined their measurements. Moreover, the recension of Eratosthenes' text
adds to the difficulties: if Eratosthenes used a stadion that was an
equivalent of 8¼ to a Roman mile, but Polybios 8⅓ and Strabo 8, this
means that all of Eratosthenes' measurements in the latter two authors
would have some amount of error: the 23,000 stadia from Borysthenes
to Meroë would be 2,788 miles using Eratosthenes' stadion, 2,761 miles
according to Polybios, and 2,875 miles according to Strabo, a difference
of 182 km. And there is no way of determining whether Strabo, when he
was quoting Polybios quoting Eratosthenes, made any conversions. One
can make some modern approximations, or perhaps note when a distance provided by Eratosthenes seems highly inaccurate, but the "cir-
cumspection" advised by Diller long ago is still essential.

Bibliography

Abel, Karlhaus. "Zone." *RE Suppl.* 14 (1974) 989–1188.

Agatharchides. *On the Erythraean Sea.* Ed. Stanley Burstein. London 1989.

Alonso-Núñez, José Miguel. "La vision de la Péninsule Ibérique chez les géographes et les historiens de l'époque hellénistique." *Sacris erudiri* 31 (1989–90) 1–8.

Aly, Wolfgang. *Strabon von Amaseia* 4: *Üntersuchungen über Text, Aufbau und Quellen der Geographika.* Bonn 1957.

Anderson, Andrew Runni. "Alexander at the Caspian Gates." *TAPA* 59 (1928) 130–63.

Aujac, Germaine. "Ératosthène, premier éditeur de textes scientifiques?" *Pallas* 24 (1977) 3–24.

———. *Ératosthène de Cyrène, le pionnier de la géographie: Sa mesure de la circonférence terrestre.* Paris 2001.

———. "Ératosthène et la geographie physique." In *Sciences exactes et sciencea appliquées à Alexandrie,* ed. Gilbert Argoud et Jean-Yves Guillaumin. Saint-Étienne 1998. 247–61.

———. "Les modes de representation du monde habité d'Aristote a Ptolémée." *AFM* 16 (1983) 13–32.

———. *Strabo et la science de son temps.* Paris 1966.

Barber, G. L. *The Historian Ephorus.* Second edition. Chicago 1993.

Bargnesi, Rodolfo. "Eratosthene, fr. IIIB38 Berger = Strabo XVI.1.21, C746: Osservazioni sulla fortuna di un paradosso idrografico." *Sungraphe* 6 (2004) 79–87.

Barnett, R. D. "Xenophon and the Wall of Media." *JHS* 83 (1963) 1–26.

Barrington Atlas of the Greek and Roman World. Ed. Richard J. A. Talbert. Princeton 2000.

Beeston, A.F.L. "Some Observations on Greek and Latin Data Relating to South Arabia." *BSOAS* 42 (1979) 7–12.

Bentham, R. M. "The Fragments of Eratosthenes." Ph.D. thesis, University of London 1948.

Berger, Ernst Hugo. *Die Geographischen Fragmente des Eratosthenes.* Leipzig 1880.

Bernhardy, Gottfried. *Eratosthenica.* Berlin 1822.

Bianchetti, Serena. "L'Eratosthene di Strabone." *Pallas* 72 (2006) 35–46.

Bickerman, E. J. *Chronology of the Ancient World.* Second edition. Ithaca 1980.

Biffi, Nicola. *Il medio oriente di Strabone: Libro XVI della Geografia.* Bari 2002.

———. *L'estremo oriente di Strabone: Libro XV della Geografia.* Bari 2005.

Blomqvist, Jerker. "Alexandrian Science: The Case of Eratosthenes." In *Ethnicity in Hellenistic Egypt,* ed. Per Bilde. Aarhus 1992. 53–69.

Boardman, John. *The Greeks Overseas: Their Early Colonies and Trade.* Fourth edition. London 1999.

Bosworth, A. B. "Alexander and the Iranians." *JHS* 100 (1980) 1–21.

———. *A Historical Commentary on Arrian's History of Alexander.* Oxford 1980–.

Bowersock, G. W. "Tylos and Tyre: Bahrain in the Graeco-Roman World." In *Bahrain through the Ages: The Archaeology,* ed. Sheikha Haya Ali Al Khalifa and Michael Rice. London 1986. 399–406.

Braund, David. *Georgia in Antiquity.* Oxford 1994.

Brown, Tuesdell S. *Onesicritus: A Study in Hellenistic Historiography.* Berkeley 1949.

Bunbury, E. H. *A History of Ancient Geography.* Second edition. London 1883.

Cary, M., and E. H. Warmington. *The Ancient Explorers.* Revised edition. Baltimore 1963.

Casson, Lionel. *The Periplus Maris Erythraei.* Princeton 1989.

———. *Ships and Seamanship in the Ancient World.* Princeton 1971.

Cimino, Massimo. "A New, Rational Endeavour for Understanding the Eratosthenes Numerical Result of the Earth Meridian Measurement." In *Compendium in Astronomy,* ed. Elias Mariolopoulos et al. Dordrecht 1982. 11–21.

Clarke, Katherine. *Between Geography and History: Hellenistic Constructions of the Roman World.* Oxford 1999.

Classen, C. J. "The Libyan God Ammon in Greece before 331 BC." *Historia* 8 (1959) 349–55.

Cohen, Getzel M. *The Hellenistic Settlements in Syria, the Red Sea Basin, and North Africa.* Berkeley 2006.

Cook, Arthur Bernard. *Zeus: A Study in Ancient Religion.* Vol. 1. Cambridge 1914.

Cordano, Federica. "Sulle fonti di Strabone per i prolegomena." *PP* 61 (2006) 401–16.

Cornell, T. J. "The Conquest of Italy." *CAH* 7.2 (second edition 1989) 351–419.

Dalby, Andrew. *Food in the Ancient World from A to Z.* London 2003.

Desanges, Jehan, and Michel Reddé. "La côte africaine du Bab el-Mandeb dans l'antiquité." In *Hommages à Jean Leclant.* Cairo 1994. 161–94.

Desautels, Jacques. "Les monts Riphées et les Hyperboréens dans le traité hippocratique des airs, des eaux et des lieux." *RÉG* 74 (1971–2) 289–96.

Dicks, D. R. *The Geographical Fragments of Hipparchus.* London 1960.

Dilke, O.A.W. "Eratosthenes." *DSB* 4 (1971) 388–93.

———. *Greek and Roman Maps.* Ithaca 1985.

Diller, Aubrey. "Agathemerus, *Sketch of Geography.*" *GRBS* 16 (1975) 59–76.

———. "The Ancient Measurements of the Earth," *Isis* 40 (1949) 6–9.

———. "Geographical Latitudes in Eratosthenes, Hipparchus and Poseidonius." *Klio* 27 (1934) 258–69.

———. *The Textual Tradition of Strabo's Geography.* Amsterdam 1975.

———. *The Tradition of the Minor Greek Geographers. American Philological Association Philological Monographs* 14. N.p., 1952.

Dorandi, Tiziano. "Philodemus." *BNP* 11 (2007) 68–73.

Dragone, Giorgio. *Eratostene e l'apogeo della scienza greca.* Bologna 1979.

———. "Introduzione allo studio della vita e della opere di Eratosthene." *Physis* 17 (1975) 41–70.

Dutka, Jacques. "Eratosthenes' Measurement of the Earth Reconsidered." *AHES* 46 (1993) 55–66.

Eder, Walter, and Johannes Renger, eds. *Herrscherchronologien der antiken Welt. (NP Supplement* 1). Stuttgart 2004.

Engels, Donald. "The Length of Eratosthenes' Stade." *AJP* 106 (1985) 298–311.

Engels, Johannes. "Die strabonische Kulturgeographie in der Tradition der antiken geographischen Schriften und ihre Bedeutung für die antike Kartographie." *OT* 4 (1998) 63–114.

Fakhry, Ahmed. *Siwa Oasis*. Cairo 1973.

Fantuzzi, Marco. "Aratus" (#4). *BNP* 1 (Leiden 2002) 955–60.

Fear, A. T. "Odysseus and Spain." *Prometheus* 18 (1992) 19–26.

Firsov, L. V. "Eratosthenes's Calculation of the Earth's Circumference and the Length of the Hellenistic Stade." *VDI* 121 (1972) 154–75.

Forbes, R. J. *Studies in Ancient Technology*. Vol. 1 Second edition. Leiden 1964.

Fraser, P. M. *Cities of Alexander the Great*. Oxford 1996.

———. "Eratosthenes of Cyrene." *PBA* 56 (1970) 175–207.

———. *Ptolemaic Alexandria*. Oxford 1972.

Frazer, James G. *Pausanias's Description of Greece*. Reprint, New York 1965.

Freeman, Philip. *Ireland and the Classical World*. Austin 2001.

Gardiner-Garden, John R. *Greek Conceptions on Inner Asian Geography and Ethnography from Ephoros to Eratosthenes*. Bloomington, Ind. 1987.

Gärtner, Hans-Armin. "Marcianus" (#1). *BNP* 8 (2006) 304–5.

Gawlikowski, Michal. "Thapsacus and Zeugma: The Crossing of the Euphrates in Antiquity." *Iraq* 58 (1996) 123–33.

Geminos. *Introduction to the Phenomena*. Ed. James Evans and J. Lennart Berggren. Princeton 2006.

Géographes Grecs. Vol. 1. Ed. Didier Marcotte. Paris 2002.

Geus, Klaus. "Eratosthenes." In *Geographie,* ed. Wolfgang Hübner. Stuttgart 2000. 75–92.

———. *Eratosthenes von Kyrene: Studien zur hellenistischen Kultur- und Wissenschaftsgeschichte*. Munich 2002.

———. "Measuring the Earth and the *Oikoumene*: Zones, Meridians, *Sphragides* and Some Other Geographical Terms Used by Eratosthenes of Cyrene." In *Space in the Roman World,* ed. Richard Talbert and Kai Brodersen. Münster 2004. 11–26.

———. "Space and Geography." In *A Companion to the Hellenistic World,* ed. Andrew Erskine. Paperback edition. Oxford 2005.

Gnoli, Gherardo. "APIANH: Postilla ad *airyō.sayana.*" *RSO* 41 (1966) 329–34.

Goldstein, Bernard R. "Eratosthenes on the 'Measurement' of the Earth." *HM* 11 (1984) 411–16.

Gottschalk, H. B. "Strato of Lampsacus." *DSB* 13 (1976) 91–5.

Grenfell, Bernard P. and Arthur S. Hunt. *The Oxyrhynchus Papyri* 10. London 1914.

———. *The Oxyrhynchus Papyri* 11. London 1915.

Gulbelkian, Edward. "The Origin and Value of the Stadion Unit Used by Eratosthenes in the Third Century B. C." *AHES* 37 (1987) 359–63.

Hamilton, J. R. "Alexander and the Aral." *CQ* n.s. 21 (1971) 106–11.

Hansman, John. "Charax and the Karkheh." *IrAnt* 7 (1967) 21–58.

Head, Barclay V. *Historia Numorum*. New and enlarged edition. London 1911.

Healy, John F. *Mining and Metallurgy in the Greek and Roman World*. London 1978.

Heidel, William Arthur. *The Frame of the Ancient Greek Maps*. New York 1937.

Hicks, Ruth Ilsley. "Egyptian Elements in Greek Mythology." *TAPA* 93 (1962) 90–108.

Hilton, John. "Azania—Some Etymological Considerations." *AClass* 35 (1992) 151–9.

Hölbl, Günther. *A History of the Ptolemaic Empire*. Tr. Tina Saavedra. London 2001.

Holt, Frank L. *Thundering Zeus: The Making of Hellenistic Bactria*. Berkeley 1999.

Hönigmann, E. "Strabon von Amaseia" (#3). *RE* 2nd ser. 4 (1931) 76–151.

Jacob, Christian. *La description de la terre habitée de Denys d'Alexandrie ou la leçon de géographie*. Paris 1990.

Jacoby, Felix. *Die Fragmente der Griechischen Historiker*. Leiden 1923–.

Jones, A.H.M. *The Cities of the Eastern Roman Provinces*. Oxford 1937.

Jones, Alexander. "Eratosthenes of Kurene." In *The Encylopedia of Ancient Natural Scientists,* ed. Paul T. Keyser and Georgia Irby-Massie. London 2008. 299–302.

Karageorghis, Vassos. *Cyprus from the Stone Age to the Romans*. London 1982.

Karttunen, Klaus. *India and the Hellenistic World*. Studia Orientalia 83. Helsinki 1997.

———. *India in Early Greek Literature*. Studia Orientalia 65. Helsinki 1989.

Keyser, Paul T. "Anaxikrates." In *The Encylopedia of Ancient Natural Scientists*, ed. Paul T. Keyser and Georgia Irby-Massie. London 2008. 74.

———. "The Geographical Work of Dikaiarchos." In *Dicaearchus of Messana*, ed. William W. Fortenbaugh and Eckhart Schütrumpf. Rutgers University Studies in Classical Humanities 10. New Brunswick 2001. 353–72.

Kidd, I. G. *Posidonius 2: The Commentary*. Cambridge 1988.

Kish, George. *A Source Book in Geography*. Cambridge, Mass. 1978.

Kleomedes. *Cleomedes' Lectures on Astronomy*. Ed. Alan C. Bowen and Robert B. Todd. Berkeley 2004.

Knaack, G. "Eratosthenes" (#4). *RE* 6 (1907) 358–88.

Korenjak, Martin. "*Italiam contra Tiberinaque longe / Ostia*: Virgil's Carthago and Eratosthenian Geography." *CQ* 54 (2004) 646–9.

Labaree, Benjamin W. "How the Greeks Sailed into the Black Sea." *AJA* 61 (1957) 29–33.

Lafond, Yves. "Cynuria" (#1). *BNP* 3 (2003) 1063.

Meiggs, Russell. *Trees and Timber in the Ancient Mediterranean World*. Oxford 1982.

Meritt, Benjamin Dean et al. *The Athenian Tribute Lists*. Vol. 1. Cambridge, Mass. 1939.

Moraux, Paul. *Les listes anciennes des ouvrages d'Aristote*. Louvain 1954.

Morel, Jean-Paul. "Les Phocéens en Occident: Certitudes et hypotheses." *PP* 21 (1966) 378–420.

Morrison, Philip, and Phylis Morrison. *The Ring of Truth: An Inquiry into How We Know What We Know*. New York 1987.

Moulay Rchid, El Mostafa. "Ératosthène. L'*Oikoumène* et la Maurusie: Géographie et symétrie." In *Mélanges Pierre Lévêque* 3: *Anthropologie et société*. Paris 1989. 269–75.

Mouraviev, S. N. "Five Ancient Testimonies in Favour of the 'Ptolemean' Transgression of the Caspian Sea." *DGT* for 1986 (1988) 235–47.

———. "Ptolemy's Map of Caucasian Albania and the Level of the Caspian Sea." *VDI* 163 (1983) 117–47.

Müller, C. *Geographi graeci minores*. Paris 1855–82.

Newton, Robert R. "The Sources of Eratosthenes' Measurement of the Earth." *QJRAS* 21 (1980) 379–87.

Pàmias, Jordi. "Dionysus and Donkeys on the Streets of Alexandria: Eratosthenes' Criticism of Ptolemaic Ideology." *HSCP* 102 (2004) 191–8.

Pearson, Lionel. *Early Ionian Historians.* Oxford 1939.

Phillips, E. D. "Odysseus in Italy." *JHS* 73 (1953) 53–67.

Plommer, Hugh. *Vitruvius and Later Roman Building Manuals.* Cambridge 1973.

Pothecary, Sarah. "Strabo, Polybios, and the Stade." *Phoenix* 49 (1995) 49–67.

Potts, Daniel T. *The Arabian Gulf in Antiquity* 2: *From Alexander the Great to the Coming of Islam.* Oxford 1990.

———."Thaj and the Location of Gerrha." *PSAS* 14 (1984) 87–91.

Pytheas of Massalia. *On the Ocean.* Ed. Christina Horst Roseman. Chicago 1994.

Queen of Sheba: Treasures from Ancient Yemen. Ed. St. John Simpson. London 2002.

Rawlins, Dennis. "Eratosthenes' Geodesy Unraveled: Was There a High-Accuracy Hellenistic Astronomy?" *Isis* 73 (1982) 259–65.

———. "The Eratosthenes-Strabo Nile Map. Is It the Earliest Surviving Instance of Spherical Cartography? Did It Supply the 5000 Stades Arc for Eratosthenes' Experiment?" *AHES* 26 (1982) 211–19.

Rickey, V. Frederick. "How Columbus Encountered America." *Mathematics Magazine* 65 (1992) 219–25.

Roller, Duane W. "Columns in Stone: Anaximandros' Conception of the World." *AntCl* 58 (1989) 185–9.

———. "Seleukos of Seleukeia." *AntCl* 74 (2005) 111–18.

———. *Through the Pillars of Herakles: Greco-Roman Exploration of the Atlantic.* London 2006.

———. *The World of Juba II and Kleopatra Selene: Royal Scholarship on Rome's African Frontier.* London 2003.

Romm, James S. *The Edges of the Earth in Ancient Thought.* Princeton 1992.

Sandars, N. K. *The Sea Peoples.* London 1978.

Seidel, Günther Carl Fridrich. *Eratosthenis geographicorum fragmenta.* Göttingen 1789.

Shcheglov, Dimitry A. "Eratosthenes' Parallel of Rhodes and the History of the System of *Climata.*" *Klio* 88 (2006) 351–9.

———. "Ptolemy's System of Seven Climata and Eratosthenes' Geography." *Geographia antiqua* 13 (2004) 21–37.

Sinclair, T. A. *Eastern Turkey: An Architectural and Archaeological Survey.* Vol. 1. London 1987.

Standish, J. F. "The Caspian Gates." *G&R* 2nd ser. 17 (1970) 17–24.

Stanley, C.V.B. "The Oasis of Siwa." *Journal of the Royal African Society* 11 (1912) 290–324.

Strabo of Amaseia. *Geographica.* Ed. Wolfgang Aly. Bonn 1968–72.

———. *Géographie* 1.1. Ed. Germaine Aujac and François Lasserre. Paris 2003.

———. *Géographie* 1.2. Ed. Germaine Aujac. Paris 2003.

———. *Géographie* 2. Ed. François Lasserre. Paris 2003.

———. *Géographie* 3. Ed. François Lasserre. Paris 2003.

———. *Géographie* 4. Ed. Raoul Baladié. Paris 2003.

———. *Géographie* 5. Ed. Raoul Baladié. Paris 2003.

———. *Géographie* 8. Ed. François Lasserre. Paris 2003.

———. *Geography.* Tr. Horace Leonard Jones. Cambridge, Mass. 1917–1932.

Syme, Ronald. *Anatolica: Studies in Strabo.* Ed. Anthony Birley. Oxford 1995.

Tarn, W. W. *Alexander the Great* 2: *Sources and Studies.* Cambridge 1948.

Tarn, W. W. *The Greeks in Bactria and India*. Revised third edition. Ed. Frank Lee Holt and M.C.J. Miller. Chicago 1997.

———. "Ptolemy II and Arabia." *JEA* 15 (1929) 9–25.

Thalamas, A. *Le géographie d'Ératosthène*. Versailles 1921.

Thomson, J. Oliver. *History of Ancient Geography*. Cambridge 1948.

Traill, John S. *The Political Organization of Attica. Hesperia Supplement* 14. 1975.

Van Beek, Gus W. "Frankincense and Myrrh." *BiblArch* 23 (1960) 69–75.

Walbank, F. W. "The Geography of Polybius." *ClMed* 9 (1947) 155–82.

———. *A Historial Commentary on Polybius*. Oxford 1970–79.

———. *Polybius*. Berkeley 1972.

West, Stephanie. "'The Most Marvellous of All Seas': The Greek Encounter with the Euxine." *G&R* 50 (2003) 151–67.

———. "Notes on the Text of Lycophron." *CQ* 33 (1983) 114–35.

Winter, Erich. "Weitere Beobachtungen zur 'Grammaire du Temple' in der Griechisch-Römischen Zeit." In *Temple und Kult*, ed. Wolfgang Helck. Wiesbaden 1987. 61–76.

Xenophanes of Kolophon. *Fragments*. Ed. J. H. Lesher. Toronto, reprint 2001.

Zimmermann, Klaus. "Eratosthenes' *Chlamys*-Shaped World: A Misunderstood Metaphor." In *The Hellenistic World: New Perspectives,* ed. Daniel Ogden. London 2002. 23–40.

Index of Passages Cited

Italicized numbers are citations in ancient texts; romanized numbers are pages in this volume.

Biblical Citations

General Index

The names Strabo and Eratosthenes occur on virtually every page of the text, and their index entries are meant to isolate important points rather than be complete. Spelling variants and ethnyms are often entered in parentheses after the main toponym. In addition, there are not as many sub-headings as one might expect, because the numerous toponyms mentioned by Eratosthenes do not admit of such categorization, and the Gazetteer serves as a method of cross-reference.

Abilyx, topographical feature, 98, 202–3, 224, 242, 252
Abioi, Homeric ethnic group, 45–7
Academy, Athenian philosophical school, 8–9, 19
Achaia, district of Greece, 17, 129, 214, 232
Achaians, Homeric ethnic group, 124
Achaimenid Period, 176
Achaimenidai, Persian ethnic group, 89, 189
Acherousian Marsh, Italian locale, 45, 224, 253
Achne, Greek toponym, 104, 214, 224, 254
Aden, Arabian city, 192
Adria(tic) Gulf or Sea, 28, 49, 55, 102, 103, 104, 126, 132, 134, 224, 225, 253, 254
Aegean Sea, 48, 124, 131, 224, 228, 232, 233, 237, 238, 239, 243, 246, 247, 254
Africa, 28, 97, 118–19, 123, 200, 224; circumnavigation of, 2, 145, 152, 156, 159; East, 10, 137, 197; Horn of, 24, 169, 197; interior, 2, 8, 25, 28, 160; North, 28, 30, 201; northwest, 137, 171, 190, 201; West, 6, 20, 116, 123, 156–7, 159–60, 180, 200–1, 107
Agamemnon. See Atreides
Agatharchides of Knidos, Greek geographer, 31
Agathemeros, Greek geographer, 148, 161
Aglaos of Kyrene, father of Eratosthenes, 7–8, 268
Agraioi, Arabian tribe, 93
Agraioi or Agraieis, northwest Greek ethnic group, 105, 216
Agrippa, M., map of, 201
Aigina, Greek island, 53, 132, 224, 254
Aigisthos, Greek hero, 42
Aigyptos River, 123. See also Nile
Ailana, Ailanitic Gulf, Arabian locale, 93, 224, 256, 258
Aiolos, Homeric figure, 45, 115
Aiolos or Aiolian Islands, 224, 253

Aischylos, Greek tragedian, 8, 9, 46, 119–121
Aithalia, Italian island, 218
Aithiopia (Aithiopes, Aithiopians), 26, 42, 46–8, 56, 64–7, 84, 91, 93–6, 98, 123, 135, 178, 192, 197, 200, 224, 239, 251, 253, 255
Aithiopian Sea, 65, 224
Aitna, Sicilian mountain, 44, 117, 224, 253
Akakesion, Greek topographical feature, 46, 224, 254
Akamas, Cypriot town, 102, 225, 256
Akarnania, Greek district, 105, 225, 254
Akko, Israeli city, 243
Albania (Albanoi), Caspian region and ethnic group, 99, 206, 223, 225, 257
Albania, Balkan country, 226, 228, 229, 230
Alexandra, Greek poem, 118, 208
Alexander the Great, 8, 14, 16, 20, 22, 127, 129, 180, 198, 225, 235, 242; and aromatics, 195–6; companions of, 18, 145–6, 157, 161, 171, 177, 193–4, 204, 272; eastern expedition of, 10, 88, 90, 98, 164, 168, 181–2, 186–7, 190, 226, 231, 232, 238, 246; and geography, 6, 93–4, 112, 128, 193–4; and India, 81, 86, 179; Istros delta expedition of, 217; and manipulation of topography, 56–7, 139–40, 163, 183–4, 204, 206; as model of virtue, 29–30, 107, 219–20; Successors of, 19, 22, 27, 51, 121
Alexandria Among the Areioi, 85, 98, 181, 225, 226, 259
Alexandria Arachosia, 138, 142, 178, 225, 226
Alexandria Eschate, 204
Alexandria in (or by) Egypt, 10, 23n98, 24–5, 58, 63, 64, 66, 75–7, 80, 81, 96–7, 101, 123, 151–2, 157, 168, 170–1, 173–4, 201, 225, 242, 250, 251, 256; and Eratosthenes, 10–15, 17, 20, 27–8, 115; Library at, 10–12, 18n87, 65, 115, 125, 172, 186